T0205540

# Multimedia Systems and Applications

**Series editor**
Borko Furht, Florida Atlantic University, Boca Raton, USA

More information about this series at http://www.springer.com/series/6298

Alexis Joly • Stefanos Vrochidis • Kostas Karatzas
Ari Karppinen • Pierre Bonnet
Editors

# Multimedia Tools and Applications for Environmental & Biodiversity Informatics

Springer

*Editors*
Alexis Joly
Inria ZENITH Team
Montpellier, France

Kostas Karatzas
Aristotle University of Thessaloniki
Thessaloniki, Greece

Pierre Bonnet
CIRAD, UMR AMAP
Montpellier, France

AMAP, Univ Montpellier
CIRAD, CNRS, INRA, IRD
Montpellier, France

Stefanos Vrochidis
Centre for Research & Technology
Hellas – Information Technologies
Institute
Thessaloniki, Greece

Ari Karppinen
FMI/Atmospheric Composition
Finnish Meteorological Institute
Helsinki, Finland

Multimedia Systems and Applications
ISBN 978-3-030-09492-8      ISBN 978-3-319-76445-0   (eBook)
https://doi.org/10.1007/978-3-319-76445-0

This Springer imprint is published by the registered company Springer International Publishing AG part of Springer Nature.
The registered company address is: Gewerbestrasse 11, 6330 Cham, Switzerland

*The authors would like to dedicate this book to the citizen scientists all around the world who invest their time, capacity and knowledge to develop research projects with the scientific community. Several research results presented in this book could not have been reached without a large and massive investment of citizens: this is why the authors would like to thank, in the broad sense, the citizen scientist community.*

# Preface

The recent and rapid advancements of digital technologies have resulted in a great increase of multimedia data production worldwide. This is also the case for multimedia data that characterize our environment and the earth biodiversity and reflect their status, behavior, change as well as human interests and concerns. Such data become more and more crucial for understanding environmental issues and phenomena. Therefore, there is an increasing need for the development of advanced methods, techniques and tools for collecting, managing, analysing, understanding and modelling environmental and biodiversity data. This edited volume focuses on the last and most impactful advancements of this field. It provides important recommendations for the implementation of computational platforms dedicated to environmental monitoring or citizen science observatories. It gives innovative and detailed architectures and specifications for the development of real-time, highly scalable, detection systems. Finally, it demonstrates the effectiveness of computational intelligence approaches in the analysis and modelling of relevant data.

Montpellier, France   Alexis Joly
Montpellier, France   Pierre Bonnet
Thessaloniki, Greece   Stefanos Vrochidis
Thessaloniki, Greece   Kostas Karatzas
Helsinki, Finland   Ari Karppinen

# Contents

# Contributors

**Alexandra Albert**  University of Manchester, Manchester, UK

**Georgia Alexandri**  Democritus University of Thrace, Xanthi, Greece

**Marius Bartcus**  DYNI Team, DYNI, Aix Marseille Univ, Université de Toulon, CNRS, LIS, Marseille, France

**Pierre Bonnet**  CIRAD, UMR AMAP, Montpellier, France

AMAP, Univ Montpellier, CIRAD, CNRS, INRA, IRD, Montpellier, France

**Thomas Le Bourgeois**  CIRAD, UMR AMAP, Montpellier, France

**Christophe Botella**  INRIA Sophia-Antipolis - ZENITH team, LIRMM - UMR 5506 - CC 477, Montpellier, France

INRA, UMR AMAP, Montpellier, France

AMAP, Univ Montpellier, CIRAD, CNRS, INRA, IRD, Montpellier, France

BioSP, INRA, Site Agroparc, Avignon, France

**Jose Carranza-Rojas**  School of Computing, Costa Rica Institute of Technology, Cartago, Costa Rica

**Luigi Ceccaroni**  1000001 Labs, Barcelona, Spain

**Faicel Chamroukhi**  LMNO UMR CNRS, Statistics and Data Science, University of Caen, Caen, France

**Bernat Claramunt**  CREAF, Edifici Ciéncies, Autonomous University of Barcelona (UAB), Bellaterra, Catalonia

Ecology Unit (BABVE), Autonomous University of Barcelona (UAB), Bellaterra, Catalonia

**Olha Danylo**  International Institute for Applied Systems Analysis (IIASA), Laxenburg, Austria

**Balasubramanian Dhandapani** French Institute of Pondicherry, UMIFRE 21 CNRS-MAEE, Pondicherry, India

**Mirjam Fredriksen** NILU – Norwegian Institute for Air Research, Kjeller, Norway

**Akio Fujiwara** GSALS, The University of Tokyo, Tokyo, Japan

**Aristeidis K. Georgoulias** Democritus University of Thrace, Xanthi, Greece

**Hervé Glotin** DYNI Team, DYNI, Aix Marseille Univ, Université de Toulon, CNRS, LIS, Marseille, France

**Hervé Goëau** CIRAD, UMR AMAP, Montpellier, France

AMAP, Univ Montpellier, CIRAD, CNRS, INRA, IRD, Montpellier, France

**Margaret Gold** National History Museum London, London, UK

**Yaela Golumbic** TECHNION – Israel Institute of Technology, Technion City, Haifa, Israel

**Pierre Grard** CIRAD, Nairobi, Kenya

**Muki Haklay** Extreme Citizen Science (ExCiteS), University College London, London, UK

**Siang Thye Hang** Toyohashi University of Technology, Toyohashi, Japan

**Philippe Jauzein** AgroParisTech UFR Ecologie Adaptations Interactions, Paris, France

**Alexis Joly** Inria ZENITH Team, Montpellier, France

**Kostas Karatzas** Aristotle University of Thessaloniki, Thessaloniki, Greece

**Ari Karppinen** FMI/Atmospheric Composition, Finnish Meteorological Institute, Helsinki, Finland

**Hill Hiroki Kobayashi** CSIS, The University of Tokyo, Tokyo, Japan

**Mike Kobernus** NILU – Norwegian Institute for Air Research, Kjeller, Norway

**Yiannis Kompatsiaris** Centre for Research and Technology Hellas – Information Technologies Institute, Thessalonki, Greece

**Renzo Kottmann** Max Planck Institute for Marine Microbiology, Bremen, Germany

**Konstantinos Kourtidis** Democritus University of Thrace, Xanthi, Greece

**Hiromi Kudo** CSIS, The University of Tokyo, Tokyo, Japan

**Christopher Kyba** GFZ German Research Centre for Geosciences, Potsdam, Germany

**Mario Lasseck** Museum fuer Naturkunde Berlin, Leibniz Institute for Evolution and Biodiversity Science, Berlin, Germany

**Hai-Ying Liu** NILU – Norwegian Institute for Air Research, Kjeller, Norway

**Soledad Luna** European Citizen Science Association (ECSA), Institute of Forest Growth and Computer Science, Technische Universität Dresden, Dresden, Germany

Nazca Institute for Marine Research, Quito, Ecuador

**Valéry Malécot** IRHS, Agrocampus-Ouest, Rennes, France

INRA, Université d'Angers, Angers, France

**Erick Mata-Montero** School of Computing, Costa Rica Institute of Technology, Cartago, Costa Rica

**Jean-Claude Melet** Independent Botanist

**Pascal Monestiez** BioSP, INRA, Site Agroparc, Avignon, France

**Anastasia Moumtzidou** Centre for Research and Technology Hellas – Information Technologies Institute, Thessalonki, Greece

**François Munoz** Université Grenoble Alpes, Saint-Martin-d'Hères, France

**Kazuhiko Nakamura** CSIS, The University of Tokyo, Tokyo, Japan

**Symeon Papadopoulos** Centre for Research and Technology Hellas – Information Technologies Institute, Thessalonki, Greece

**Jaume Piera** Institute of Marine Sciences (ICM-CSIC), Barcelona, Spain

**Marion Poupard** AMU, University of Toulon, UMR CNRS LIS, Marseille, France

**Antonella Radicchi** Technical University Berlin, Berlin, Germany

**Prabhakar Rajagopal** Strand Life Sciences, Bangalore, India

**Johanna Robinson** JSI – Jožef Stefan Institute, Ljubljana, Slovenia

**Vincent Roger** DYNI Team, LIS UMR CNRS 7020, AMU, University of Toulon, Marseille, France

**Kaoru Saito** CSIS, GSFS, GSALS, The University of Tokyo, Tokyo, Japan

**Sven Schade** European Commission, Joint Research Centre (JRC), Unit B06-Digital Economy, Ispra, Italy

**Kaoru Sezaki** CSIS, The University of Tokyo, Tokyo, Japan

**Daisuké Shimotoku** GSFS, The University of Tokyo, Tokyo, Japan

**Eleftherios Spyromitros-Xioufis** Centre for Research and Technology Hellas – Information Technologies Institute, Thessalonki, Greece

**Ulrike Sturm** Museum für Naturkunde Berlin, Leibniz Institute for Evolution and Biodiversity Science, Berlin, Germany

**Milan Šulc** Czech Technical University in Prague, Prague, Czech Republic

**Thomas Vattakaven** Strand Life Sciences, Bangalore, India

**Stefanos Vrochidis** Centre for Research & Technology Hellas – Information Technologies Institute, Thessaloniki, Greece

**Christian You** Société Botanique Centre Ouest, Nercillac, France

# Acronyms

# Chapter 1
# Introduction

Alexis Joly, Pierre Bonnet, Stefanos Vrochidis, Kostas Karatzas,
and Ari Karppinen

**Abstract**  The recent and rapid advancements of digital technologies, as well as the progress of digital cameras and other various connected objects have resulted in a great increase of multimedia data production worldwide. Such data becomes more and more crucial for understanding environmental issues and phenomena, such as the greenhouse effect, global warming and biodiversity loss. Therefore, there is an increasing need for the development of advanced methods, techniques and tools for collecting, managing, analyzing and understanding environmental and biodiversity data. The goal of this introductory chapter is to give a global picture of that domain and to overview the research works presented in this book.

A. Joly (✉)
Inria ZENITH Team, Montpellier, France
e-mail: alexis.joly@inria.fr

P. Bonnet
CIRAD, UMR AMAP, Montpellier, France

AMAP, Univ Montpellier, CIRAD, CNRS, INRA, IRD, Montpellier, France
e-mail: pierre.bonnet@cirad.fr

S. Vrochidis
Centre for Research & Technology Hellas - Information Technologies Institute, Thessaloniki, Greece
e-mail: stefanos@iti.gr

K. Karatzas
Aristotle University of Thessaloniki, Thessaloniki, Greece
e-mail: kkara@eng.auth.gr

A. Karppinen
FMI/Atmospheric Composition, Finnish Meteorological Institute, Helsinki, Finland
e-mail: ari.karppinen@fmi.fi

© Springer International Publishing AG, part of Springer Nature 2018
A. Joly et al. (eds.), *Multimedia Tools and Applications for Environmental & Biodiversity Informatics*, Multimedia Systems and Applications,
https://doi.org/10.1007/978-3-319-76445-0_1

1

## 1.1 From Field Studies to Multimedia Data Flows

During centuries, the study of earth biodiversity and environment mostly relied on field studies conducted by highly skilled experts from museums and universities who travelled the world to collect samples. Only recently did we see the emergence of alternative observation practices. First of all, the success of social networks and citizen sciences has fostered the emergence of large and structured communities of observers (e.g. e-bird, zooniverse, iNaturalist, iSpot, Tela Botanica, Luftdaten, Citizen Weather Observing Program, etc.). Citizens have become increasingly aware of the importance of watching biodiversity and ecosystems in particular because of their impact on human health and well-being (e.g. allergies, food safety, water quality, landscape management, etc.). As a consequence, a very large number of citizen science projects have been launched all over the world during the last decade. To inventor them and facilitate experience sharing, dedicated portals have been developed at the international and national level (such as SciStarter,[1] EU BON,[2] or NatureFrance[3]).

In parallel to the increased engagement of human observers, the development of new acquisition devices also boosted the emergence of alternative observation practices. Indeed, the quality, capacity and diversity of connected objects have progressed dramatically during the last decade. These new devices can produce, store and transmit large volumes of data acquired automatically or semi-automatically. Smart phones in particular allow to dramatically increase the number of people in capacity to produce simple but very useful information for environmental and biodiversity monitoring. Besides, fully autonomous audio-visual sensors start to be installed all over the world such as underwater cameras [4], camera traps [6], fisheye cameras [5], bio-acoustic recorders [1] or hydrophones [2]). All these devices produce huge data streams that are clearly under exploited today because of the lack of efficient tools to process them. A last important source of data that has emerged recently is the digitization of old materials such as natural history collections. This digitization process has been largely accelerated these last years thanks to newly developed equipment and consistent dedicated funding. For instance, biological specimens have been massively digitized in recent years [3] resulting in millions of digital records.

Based on the study of various initiatives that produce and exploit such multimedia data flows for biodiversity and environment monitoring, we can draw some stable patterns, as illustrated in Fig. 6.1. As a first necessary step, environmental records or biodiversity observations have to be integrated in accessible databases. To facilitate this step, common formats and international standards are being developed by the scientific community. For instance, the Taxonomic Databases Working Group (also

---

[1] https://scistarter.com/finder.

[2] http://biodiversity.eubon.eu/web/citizen-science/view-all.

[3] http://www.naturefrance.fr/sciences-participatives.

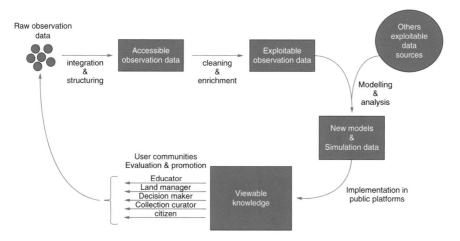

**Fig. 1.1** Biodiversity and environment monitoring dataflow

known as Biodiversity Information Standards), is in charge of the development of standard formats for the exchange of biological/biodiversity data. As an other example, the Unidata Program Center is a diverse community of education and research institutions with the common goal of sharing geoscience data and the tools to access and visualize that data. Once made accessible, raw observation data often needs to be cleaned/or and enriched before being exploited. For instance, degraded, inconsistent or duplicated data samples might need to be filtered out. Or some measurements/properties might need to be extracted from raw audio-visual content (e.g. categorical names, objects count, color attributes, visible surfaces, etc.). Once made exploitable, observation data can be used for modelling and simulation. More and more often, this is done by combining several observation data sources (e.g. environmental data and species occurrences are combined for estimating species distribution models in ecology). The resulting models and simulation data can then be integrated in publicly available visualization tools. This allows different communities of end users (experts and non experts) to appropriate the knowledge produced and to act accordingly. Then, the conclusions drawn from that knowledge might be used for the planning of new observation campaigns (Fig. 1.1).

## 1.2   Book Content Overview

Figure 1.2 presents an overview of the typical processing pipeline to be implemented when setting up an end-to-end environmental monitoring system. As illustrated by the Gantt chart below, the chapters of this book relate to one or several steps of this pipeline and were sorted based on their location within this pipeline.

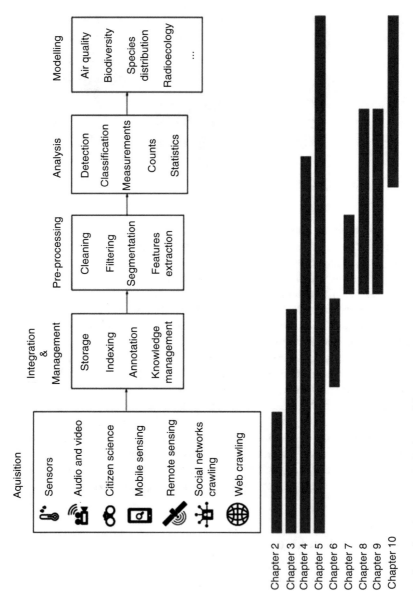

**Fig. 1.2** Biodiversity and environment monitoring workflow

Accordingly, the two first chapters are focused on data acquisition, in the specific context of citizen science and participatory sensing. Chapter 2 first highlights that the ubiquity of powerful mobile devices has brought about a proliferation of mobile phone based citizen science projects and suggests that there is a lack of systematic knowledge exchange on the development of mobile apps and web platforms for citizen science. The authors aim at filling this gap, specifically by surveying the key considerations for the effective development of mobile applications and web platforms. As a concrete illustration of this research field, Chap. 3 presents an existing platform that was developed within a European project for the understanding and implementing of citizens' observatory on air monitoring. It presents a number of tools from the user and the developer perspective that together provide the foundation of a citizens' observatory.

Chapters 4 and 5 are rather focused on the presentation of automated monitoring systems. Chapter 4 presents a real-time streaming and detection system for bio-acoustic ecological studies after the Fukushima accident. Audio recordings were continuously collected within the exclusion (i.e., difficult-to-return-to) zone located 10 km from the Fukushima Daiichi Nuclear Power Plant in the Omaru District (Namie, Fukushima, Japan). The authors describe the potential of this live stream of sound data from an unmanned remote station as well as the methodologies by which they processed these recordings as eco-acoustic indexes for demonstrating possible correlations between biodiversity variation and preexisting radioecology observations. As a last remarkable monitoring system, Chap. 5 presents an end-to-end framework for automatic air quality evaluation from raw images collected from social media. The framework includes three main parts: (1) a timestamped and geo-tagged data collection mechanism, (2) a sky detection/localization system that aims at retaining only images containing sky and highlighting only sky related pixels and (3) an air quality evaluation algorithm based on sky related pixel intensities evaluation (only applied to sky related pixels identified by the previous module).

As illustrated on Fig. 1.2, the remaining chapters are more focused on specific steps of the overall processing pipeline. First of all, Chap. 6 deals with data integration and knowledge representation issues. In particular, it discusses the challenges related to the aggregation and the exploitation of traits in an information system. Traits are a crucial information for structuring and accessing to the knowledge about any living organism (e.g. morphology, taxonomy, functional role, habitat, ecological interactions, etc.). The paper addresses several challenging tasks such as aggregating such information, structuring it for a better navigation and then exploiting it for new applications. Regarding the two first steps, the paper introduces a concrete platform (BIP, standing for Biodiversity Informatics Platform) dedicated to the integration of species information and specimen observations from both experts and citizens.

Further, Chaps. 7, 8 and 9 focus on the pre-processing and analysis of audio-visual signals in the context of biodiversity monitoring. Chapter 7 first addresses one of the most crucial pre-processing step for the analysis of bio-acoustic signal, i.e. its segmentation. Therefore, the authors propose to rely on the Hierarchical Dirichlet Process for Hidden Markov Model (HDP-HMM). Using this model is well justified because it solves the major problem of fixing the number of states (song units) that reveal the diversity of sounds emitted for animal communication at the intra-level species and at an inter-level species. This study demonstrates new insights for unsupervised analysis of complex soundscapes and illustrates their potential of chunking non-human animal signals into structured units. This can yield to new representations of the calls of a target species, but also to the structuring of inter-species calls. Chapter 8 rather deals with the automated identification of plants in images. This challenging task improved considerably in the last few years, in particular thanks to the recent advances in deep learning. The central question addressed within the chapter is to know how far such automated systems are from the human expertise. Indeed, even the best experts are sometimes confused and/or disagree between each others when validating visual or audio observations of living organism. A picture or a sound actually contains only a partial information that is usually not sufficient to determine the right species with certainty. Quantifying this uncertainty and comparing it to the performance of automated systems is of high interest for both computer scientists and expert naturalists. Chapter 9 also addresses the problem of identifying plant groups in images but it differs in two main points. First of all, it deals with digitized herbarium specimens rather than in-the-field photographs as in the previous chapter. The problem is more challenging in that the preservation process of pressing and drying plants for herbarium purposes might cause important changes and loss of information. Furthermore, the paper considers the problem of predicting the genus and the family of the samples and not only the species taxonomic level. They introduce several deep learning architectures that are compared through a large-scale comparative study involving thousand of species.

Last but not least, Chap. is the only one that is centrally focused on modelling issues. It shows that multimedia technologies can have a great contribution at this level of analysis. More precisely, it proposes a deep learning approach to species distribution modelling (SDM) that is a central problem in ecology. Given a set of species occurrence, the aim is to infer its spatial distribution over a given territory. Because of the limited number of occurrences of specimens, this is usually achieved through environmental niche modeling approaches, i.e. by predicting the distribution in the geographic space on the basis of a mathematical representation of their distribution in environmental space (temperature, precipitation, soil type, land cover, etc.). This study is the first one evaluating the potential of the deep learning approach for this problem. It shows that deep neural network and convolutional neural networks (CNN) in particular clearly outperform classical approaches used in ecological studies, such as Maxent. This result is promising for future ecological studies developed in collaboration with naturalists expert. Actually, many ecological studies are based on models that do not take into account spatial patterns in environmental variables. On the contrary, CNN can capture extra information contained in spatial patterns of environmental variables in order to surpass other classical approaches.

# References

1. Wiggins, S. (2003). Autonomous acoustic recording packages (ARPs) for long-term monitoring of whale sounds. Marine Technology Society Journal, 37(2), 13–22.
2. Fox, C. G., Matsumoto, H., & Lau, T. K. A. (2001). Monitoring Pacific Ocean seismicity from an autonomous hydrophone array. Journal of Geophysical Research: Solid Earth, 106(B3), 4183–4206.
3. Beaman, R. S., & Cellinese, N. (2012). Mass digitization of scientific collections: New opportunities to transform the use of biological specimens and underwrite biodiversity science. ZooKeys, (209), 7.
4. Fisher, R.B., Chen-Burger, Y.H., Giordano, D., Hardman, L., Lin, F.P.: Fish4Knowledge: Collecting and Analyzing Massive Coral Reef Fish Video Data, vol. 104. Springer (2016)
5. Nagai, S., Maeda, T., Gamo, M., Muraoka, H., Suzuki, R., Nasahara, K.N.: Using digital camera images to detect canopy condition of deciduous broad-leaved trees. Plant Ecology & Diversity 4(1), 79–89 (2011)
6. Silver, S.C., Ostro, L.E., Marsh, L.K., Maffei, L., Noss, A.J., Kelly, M.J., Wallace, R.B., Gómez, H., Ayala, G.: The use of camera traps for estimating jaguar panthera onca abundance and density using capture/recapture analysis. Oryx 38(2), 148–154 (2004)

# Chapter 2
# Developing Mobile Applications for Environmental and Biodiversity Citizen Science: Considerations and Recommendations

Soledad Luna, Margaret Gold, Alexandra Albert, Luigi Ceccaroni, Bernat Claramunt, Olha Danylo, Muki Haklay, Renzo Kottmann, Christopher Kyba, Jaume Piera, Antonella Radicchi, Sven Schade, and Ulrike Sturm

**Abstract** The functionality available on modern 'smartphone' mobile devices, along with mobile application software and access to the mobile web, have opened up a wide range of ways for volunteers to participate in environmental and biodiversity research by contributing wildlife and environmental observations, geospatial information, and other context-specific and time-bound data. This has brought about an increasing number of mobile phone based citizen science projects that are designed to access these device features (such as the camera, the microphone, and GPS location data), as well as to reach different user groups, over different project durations, and with different aims and goals. In this chapter we outline a number of key considerations when designing and developing mobile applications for citizen

S. Luna (✉)
European Citizen Science Association (ECSA), Institute of Forest Growth and Computer Science, Technische Universität, Dresden, Germany

Nazca Institute for Marine Research, Quito, Ecuador
e-mail: sluna@institutonazca.org

M. Gold
National History Museum London, London, UK

A. Albert
University of Manchester, Manchester, UK

L. Ceccaroni
1000001 Labs, Barcelona, Spain

B. Claramunt
CREAF, Edifici Ciéncies, Autonomous University of Barcelona (UAB), Bellaterra, Catalonia

Ecology Unit (BABVE), Autonomous University of Barcelona (UAB), Bellaterra, Catalonia

© Springer International Publishing AG, part of Springer Nature 2018      9
A. Joly et al. (eds.), *Multimedia Tools and Applications for Environmental & Biodiversity Informatics*, Multimedia Systems and Applications,
https://doi.org/10.1007/978-3-319-76445-0_2

science, with regard to (1) Interoperability. The factors that influence the usability of the mobile application are covered in both (2) Participant Centred Design and Agile Development, and (3) User Interface and Experience Design. Finally, the factors that influence sustained engagement in the project are covered in (4) Motivational Factors for Participation.

## 2.1 Introduction

Many modern day citizen science projects are powered by mobile and web technologies, which enable the general public to take part in research and contribute to scientific knowledge around the globe [1–4]. The nature of these apps and web platforms vary almost as greatly as the underlying science [5–8], and so do the ways in which participants interact with their mobile devices and with other participants.

A systematic search of citizen science projects conducted by Pocock et al. [9] found 509 projects that fit the definition of environmental and ecological citizen science, of which 77% were focused on biodiversity rather than the abiotic environment, and 93% invited volunteers purely to contribute data, as opposed to taking a collaborative or co-created project approach. Of those 509 projects, 142 requested the submission of a photo as the core data type, 62 projects were found to require a smartphone for their execution, and 5 made use of SMS messaging.

Mobile applications to support environment and biodiversity monitoring are most commonly used to record the presence and location of native and invasive species, to date and geo-reference different biological events such as reproduction, and to identify patterns of land or seabed cover [7, 10].

O. Danylo
International Institute for Applied Systems Analysis (IIASA), Laxenburg, Austria

M. Haklay
Extreme Citizen Science (ExCiteS), University College London, London, UK

R. Kottmann
Max Planck Institute for Marine Microbiology, Bremen, Germany

C. Kyba
GFZ German Research Centre for Geosciences, Potsdam, Germany

J. Piera
Institute of Marine Sciences (ICM-CSIC), Barcelona, Spain

A. Radicchi
Technical University Berlin, Berlin, Germany

S. Schade
European Commission, Joint Research Centre (JRC), Unit B06-Digital Economy, Ispra, Italy

U. Sturm
Museum für Naturkunde Berlin, Leibniz Institute for Evolution and Biodiversity Science, Berlin, Germany

In order to be successful, most citizen science projects require a sufficient number of participants over an extended period of time. Furthermore, the ability to meet the goals of the project will depend on the usability of the mobile application from the user's perspective, its effectiveness in carrying out its purpose from the research perspective, and whether the project itself is able to communicate and disseminate the apps and web platform to the public and sustain their engagement for a sufficiently long period of time.

Each of these factors present a range of unique challenges and pitfalls to be taken into consideration when designing and building a mobile app and web platform.

To the best of our knowledge, there is no systematic exchange of experience, knowledge, and gaps-to-be-addressed for the development of such mobile apps and web platforms for citizen science. We therefore asked citizen science practitioners and project managers to identify key considerations for the effective development of mobile applications and their adherent web platforms. We did this by way of two workshops on the topic of "Defining Principles for Apps and Platform Development for Citizen Science" that were held in Berlin on the 13th and 14th of December, 2016, and in Gothenburg on the 25–27th of April, 2017, in which a total of 75 practitioners took part in person or online.

This chapter summarises the outcomes of these workshops and online contributions, wherein we highlight a number of considerations for the designing, building and development of effective and sustainable applications for environmental and biodiversity mobile-based citizen science projects. The definitions of the terminology that we use in this chapter were discussed during the workshops and agreed upon among participants (Fig. 2.1).

In this chapter, the factors that we deem important to consider and plan for at the outset of the design and build phase are described in (1) Interoperability. The factors that influence the usability of the mobile application are covered in both (2) Participant Centred Design and Agile Development, and (3) User Interface and Experience Design. Finally, the factors that influence sustained engagement in the project are covered in (4) Motivational Factors for Participation.

## 2.2 Interoperability

Interoperability can refer to the ability of humans and machines to pass information between each other via shared terminology and semantic metadata [11], or to the ability of computer systems or software to exchange information between each other and make use of that information [12].

In this chapter we focus on systems interoperability, but recognise that shared terminology (which can range as widely as citizen science, crowdsourcing, citizen engagement, public participation in science, voluntary mapping, and more) between practitioners in the field, and between participants and project initiators is equally vital. Unifying these terms greatly assists with the sharing of knowledge and emerging best practice amongst those developing apps for citizen science [13].

---

**App:** "a self-contained program or piece of software designed to fulfil a particular purpose. It is an application, especially as downloaded by a user to a mobile device." (Oxford English Dictionary)

**Citizen science:** the collection and analysis of data relating to the natural world by members of the general public, in partnership with scientists and researchers, in aid of scientific research.

**Citizen science participant / citizen scientist:** a member of the general public who does not necessarily have scientific training, who takes part in a citizen science project on a voluntary basis.

**Citizen science practitioner:** anyone involved in the active development of citizen science, e.g. researcher/scientist, project manager, technical person, science communication professionals, educators, volunteer contributor, authorities, institutions, NGOs, etc.

**Data:** information collected in an electronic format that can be stored and used by a computer.

**Forking / Software Fork:** to develop a new variant of the software on the same code basis but often with an entirely new branding.

**Platform:** a (computing) platform is a technical framework on which one or more applications may be run and where data are kept. For the purposes of user interaction (UI) and user experience (UX), the term "website" instead of platform will be used.

**Portal:** web-site providing access or links to other sites. Here, especially pointing to apps, platforms, projects etc.

---

**Fig. 2.1** List of terminology used in this chapter. Definitions were agreed upon among workshop participants

Semantics is even more important in conversations between humans and machines, or between machines [14].

## 2.2.1 Data and Metadata Standards

A common or interoperable structure and representation for data and metadata is needed in order to ensure that data can be shared and aggregated with other current and future projects. Such (meta)data includes information about citizen science projects, datasets, tools used (software, hardware, apps, instruments, sensors), and (domain specific) observations made by participants. Different organisations use different software solutions to organize knowledge gathered in or used by citizen science projects. These solutions can facilitate or impede interoperability. A number of existing data standards and metadata schemas that are used in citizen science projects are presented in Appendix A. More schemas and their documentation can, for example, be found at schema.org.

## 2.2.2 Data Sharing and Access

In order to aid data sharing across scientific applications, research projects and academic papers, a universally unique identifier (UUID) is assigned to each observation or data point in order to avoid duplication in global databases (such as the Global Biodiversity Information Facility—GBIF[1]), and to be uniquely identified without significant central coordination.

After following a data model or schema as described in Appendix A, we recommend that the data is made available to other researchers via a data service, most usually on the web via an Application Programming Interface (API). A range of standards are available for this purpose. Some are more complex and have a high learning curve but capture a rich set of diverse use cases (thus allowing for a high degree of interoperability). Examples of more lightweight alternatives come from within the Web Services of the Open Geospatial Consortium (OGC) and include the Web Feature Service[2] or the Sensor Web Enablement suite of standards[3] These cover less rich structures but are more easy to learn and apply. Another example outside the OGC is the recently revised Semantic Sensor Network (SSN) ontology of the World Wide Web Consortium (W3C),[4] which may for example be queried via SPARQL,[5] a dedicated language to query information sources following the Linked Data paradigm [15].

## 2.2.3 Data Sharing with Participants

Two very important principles for any citizen science project, as stated in the ECSA 10 Principles of Citizen Science,[6] are (1) that citizen scientists receive feedback from the project in terms of how their data are being used and what the research, policy or societal outcomes are (Principle Four), and (2) that project data and metadata are made publicly available and where possible, results are published in an open access format (Principle Seven).

It is therefore vital that project initiators plan for the sharing of both data and outcomes when establishing the project communication channels, with the participants of the project in mind, not just fellow researchers and scientists in the relevant fields. Both data and outcomes should be presented in a format that is easy for participants to navigate and understand.

---

[1] gbif.org.

[2] opengeospatial.org/standards/wfs.

[3] opengeospatial.org/ogc/markets-technologies/swe.

[4] w3.org/TR/vocab-ssn/.

[5] w3.org/TR/vocab-ssn/.

[6] ecsa.citizen-science.net/sites/default/files/ecsa_ten_principles_of_citizen_science.pdf.

Pocock et al. [9] found that mass participation projects were more likely to present their data dynamically (e.g. in real time rather than in summary reports) and in an elaborate format, whereas simple projects and entirely computer-based projects were less likely to make data available to view and download at a high resolution (e.g. full dataset, rather than data summaries or reports).

Indirect ways to make data and metadata available to participants are overarching portals such as EMODnet,[7] the GEOSS portal,[8] or GBIF[9]—all of which provide full and open access to observation data sets.

## *2.2.4   Open Data and Licensing*

Open data licenses, such as those from the Creative Commons shown in Appendix B allow for the reuse of data, and can take different countries' regulations into account when a project is global or multi-national in scope. Among the Creative Commons licenses, GBIF recommends the use of "No rights reserved" (CC0), CC-BY, or CC-BY-NC. Other formats such as the Open Data Commons licenses are particularly well suited for data licensing in a citizen science context, as pointed out by Groom et al. [16], because the Creative Commons licenses were designed with creative content in mind.

## *2.2.5   Software Reuse*

Existing apps can be reused for biodiversity monitoring when requiring little customization, avoiding the need to create a new application from scratch. Examples that offer an excellent solution are iNaturalist, Natusfera or iSpot (see Appendix C for more examples).

Another option is to use platforms that have been built to support multiple mobile-based projects, such as the Spotteron[10] platform service for fully-customisable smartphone applications for citizen science, or the Epicollect 5[11] platform for creating bespoke mobile questionnaires with data mapping on a hosted website.

Yet one of the challenges for reusability remains the aspect of discovery. So far, no comprehensive repository of reusable mobile applications for citizen science exists. However, there are several global and national citizen science project

---

[7] emodnet.eu.

[8] earthobservations.org/geoss.php and geoportal.org/.

[9] gbif.org/ipt.

[10] spotteron.net.

[11] five.epicollect.net.

directories that are a useful source of information about the full range of projects and the tools that they use, such as:

- SciStarter[12]
- Citizen Science Central[13]
- CitSci[14]
- Scientific American[15]
- UK Environmental Observation Framework[16]
- The Federal Crowdsourcing and Citizen Science Catalog[17]
- Biocollect-Atlas of Living Australia[18]
- Bürger schaffen Wissen[19]
- Citizen Science Austria[20]
- Schweiz Forscht[21]
- Iedereen een Wetenschapper[22]

## 2.2.6  Software Reusability

Since open source apps and platforms permit a higher level of customization and take advantage of a well-developed code base, it is valuable to open and share the code on a public repository such as GitHub. To maximise reuse, a good repository will include code documentation, requirement specifications, design specifications, test scenarios and results, lessons-learned documentation, and any other materials that will make it easy to 'fork' the code for a new project.

For example, the application Natusfera used a copy of the source code from iNaturalist and started an independent development on it, creating a distinct and separate piece of software. Therefore, Natusfera is a fork of iNaturalist, with its own database, look-and-feel, and special functionalities such as enabling project hierarchies.

---

[12] scistarter.com.

[13] birds.cornell.edu/citscitoolkit/projects.

[14] citsci.org.

[15] scientificamerican.com/citizen-science/.

[16] ukeof.org.uk/catalogue.

[17] ccsinventory.wilsoncenter.org.

[18] biocollect.ala.org.au.

[19] buergerschaffenwissen.de.

[20] citizen-science.at.

[21] schweiz-forscht.ch.

[22] iedereenwetenschapper.nl.

Additionally, forking open code facilitates the interoperability with the original database, and contributes to the growth of the two platforms by sharing improvements to the underlying base code.

### 2.2.7  Data Management and Data Privacy

Data Management has become one of the central challenges to emerge with the growth of citizen science projects [17, 18]. One important aspect of this is data privacy. Although scientists are naturally inclined to capture as much data as possible, including for the community of participants, it is better practice to capture as little personal data as possible the only minimum needs of the project.

Additionally, participants have to be provided with the means to indicate how their data may or not be used or shared, and it is generally considered best practice for this to be provided as an opt-in, rather than an opt-out (see for example the UK Information Commissioner's Office Guidelines for Small Businesses collecting information about their customers[23]). For example, if data points will be shown on a publicly available map, it is critical that the participants understand and consent to this, as observations taken and shared may reveal home locations or other personal details, even if their user ID is anonymized.

Moreover, project managers are responsible for secure data transmission and storage. Personal data have to be deleted as soon as possible if they are no longer needed to meet the objectives of the project. In other cases, data can be obfuscated using reliable methods that keep the data meaningful, but without disclosing details about the participant [19, 20].

These aspects of data management in citizen science are starting to gain attention in the literature. Bastin et al. [21] present the current state of the art regarding data management practices, schemas and tools, along with best-practice examples, and a range of open source technologies which can underpin robust and sustainable data management for citizen science. Additionally, Williams et al. [22] discuss how to sustain and maximize the impact of citizen science data.

### 2.2.8  Data Quality

One unique aspect of citizen science contributed data is data quality and the connected question of reliability. Therefore, in addition to standard data validation techniques, citizen science projects might also put additional effort in cross-validation data by comparing collected data to other sources such as remote

---

[23]ico.org.uk/media/for-organisations/documents/1584/pn_collecting_information_small_business_checklist.pdf.

sensing data [23]. Double bookkeeping approaches such as asking for pen and paper documentation of measurements in addition to mobile app based reporting could be used. Comparison to other data reveals outliers and establishes a general level of trust. Double bookkeeping allows for the identification of discrepancies in reporting, and hence the potential measurement of data issues. In addition, double bookkeeping is a fallback in case of malfunctions of mobile apps and data transmission and includes people without, or incompatible, smartphones.

### 2.2.9 Data Policy Transparency

Essential project information, such as how data is shared, should be made available to participants in a way that is completely transparent, but also removes friction in the user experience. For example, the Loss of the Night app[24] (Appendix C), had a participant contact the team asking for their data to be deleted, because this detail was buried in a "Terms and Conditions" page. This can be addressed by allowing participants to dive straight into the first project task, such as taking a photo, and providing the relevant data policy as part of the next step—such as a 'Submit Photo' button with an explanation that the photograph will be made public. This has the additional benefit of lowering barriers to participation, by facilitating the citizen scientist to get on with a project task.

## 2.3 Participant Centered Design and Agile Development

The central aim of citizen science is to involve the general public in scientific research, therefore projects are usually designed to involve as broad a range of participants as possible [4, 24, 25]. This can increase complexity in terms of the range of participants' interests, abilities and motivation [3, 26–28, 31].

Participant centered design (or user-centred design in the context of mobile apps development in general) helps reach and involve participants [29] by involving them throughout the entire process, from concept—to design—to iterative user-testing—to shared outcomes. The early involvement of participants helps unearth issues such as ergonomic factors and how to support the learning curve before final user testing takes place. It also allows the project to be structured for mutual benefit, for both the researchers and the participants', as well as ensuring a good user experience.

The development process of the app Naturblick[25] (Appendix C) is a good example of how to conduct participant centered design in citizen science. Potential participants were involved from the beginning by asking them about their interests

---

[24] verlustdernacht.de.

[25] naturblick.naturkundemuseum.berlin.

and ideas which fed into the conceptual process. During the development, iterative user-testing was conducted. The methods for the user-testing were adapted to the state of the development process, and ranged from focus groups to monitored testing situations with follow-up interviews. The issues and ideas were fed into the agile (i.e. iterative and incremental) development process, and resulted in prototypes for further testing and discussion. This process continued after releasing the app, which is crucial to the agile development process.

## 2.4   User Interface and User Experience Design

So far we have stressed the importance of taking interoperability and data management concerns into account at the outset of any new citizen science project, and of pursuing a participant-centred process throughout the design and build phase. In this section we now look more closely at the usability of the mobile application from the user's perspective. In mobile applications and web platforms developed for environmental and biodiversity citizen science, the user interface and the user experience are important factors to keep participants engaged and motivated [30, 31].

User interface design refers to what is displayed on the mobile phone screen or website, with considerations such as choosing a clear typeface, a well-contrasted and visible colour palette, effective use of images and the placement of buttons, links or arrows.

User experience design refers to how the steps to be taken are placed in a logical flow, such that the project participants are eased through each step. Design elements need to be both effective and efficient, influencing how the participants perform certain interactions, and guiding them through the steps to be taken.

### 2.4.1   Mobile Applications and Websites

It is typical for mobile phone based citizen science projects to provide both a website and a smartphone application (the app) as illustrated in Fig. 2.2. Modern smartphones are a powerful tool for data-collection in the field, enabling citizen scientists to take measurements, document and photograph their observations, record geo-location data, and easily upload these data to a shared repository.

Crowdsourced contributions via website interfaces include entering and uploading observation data that were recorded on the mobile phone, processing and analysing data, and transcribing existing data into a digital format. Mobile and web interfaces have to be designed in a way that simplifie data gathering, encourages participation by as wide a range of people as possible, and ideally increases scientific understanding as well.

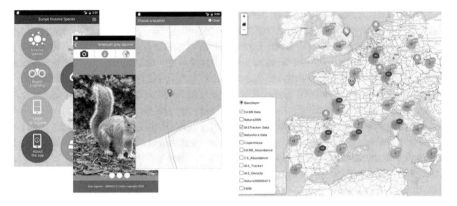

**Fig. 2.2** Examples of a mobile citizen science app (left), and a web page integrating gathered observations from this app and two others (right). Source: digitalearthlab.jrc.ec.europa.eu/app/invasive-alien-species-europe

To ensure that the interfaces are accessible to the widest possible audience, it is important to use open web standards such as the HTML5 markup language, which is ideal for cross-platform mobile applications, and to ensure that the interface and API are RESTful—i.e. based on representational state transfer (REST) technology, an architectural style and approach to communications used in web services that ensures operability, robustness, and scalability [32].

### 2.4.2 UI/UX Best Practice in the Software Development and Design Literature

A brief internet search using the software development industry's shorthand of UI/UX (i.e. user interface and user experience) displays digital magazines showcasing new design trends and design patterns (e.g. Hongkiat.com, and theUXReview.co.uk), digital magazines showcasing new design element trends and winning designs (e.g. SmashingMagazine.com and UXmag.com), and indispensable tech-know-how reference books such as *Effective UI: The Art of Building Great User Experience in Software* [33] and *Mobile First* [34].

The mobile operating system providers also create highly useful guides for developing native applications for their platforms, such as the "Think with Google" series on *Principles of Mobile App Design*[26] and the Android developer centre *Design Guides.*[27]

---

[26]thinkwithgoogle.com/marketing-resources/experience-design/principles-of-mobile-app-design-introduction/.

[27]developer.android.com/design/index.html.

The primary general principle touted by most practitioners of UI/UX design is to strip the design back to the most simple functionality possible. The concept of 'Minimum Viable Product' from Lean Startup thinking refers to "the version of a new product which allows a team to collect the maximum amount of validated learning about customers with the least effort" [35], and the 'Simplicity Principle' from design thinking which states that: "the design should make simple, common tasks easy, communicating clearly and simply in the user's own language, and providing good shortcuts that are meaningfully related to longer procedures" [36].

Using existing UX patterns, such as the 'hamburger' three stripes icon that indicates a menu that can be opened up for further navigation, will help project participants to feel confident that they can find their way around the app, following familiar conventions.

Another general rule of thumb for anything digital is to reduce the number of actions, or clicks, as much as possible, since user-testing consistently shows drop-off of usage with each step to be taken. This is sometimes known as the 'Three Click Rule' [37]. A further important consideration, still frequently overlooked, is to take accessibility into account by following the Web Content Accessibility Guidelines (WCAG).[28]

The key recommendation here is that sufficient time be spent perusing these useful guides to best practice in UI/UX design before embarking on the design and implementation phase of any application.

### 2.4.3 UI/UX Considerations Specific to Citizen Science Projects

Citizen science projects that propose to reach out to audiences with low science capital [38, 39], should conduct user-profiling to understand who is likely to use the mobile application, and in what context. For example, if the goal of the project is to reach out to school-aged children, use of language should be kept simple, and images could be used to illustrate next steps. The UCL ExCiteS group has developed the Sapelli platform[29] for mobile data collection and data sharing in Citizen Science projects where the participating group are non-literate or illiterate, with little or no prior information and communications technology (ICT) experience.

Usability testing and contextual research are essential practice in this regard, allowing the project initiators to observe real users interacting with the mobile application to catch potential design improvements. An excellent case study of user testing amongst both citizen science practitioners and participants for the Creek Watch monitoring app is contained in Kim et al. [42].

---

[28] w3.org/WAI/intro/wcag.

[29] sapelli.org.

"Several participants (ten environmental scientists in the City of San Jose Environmental Services Water Resources Department in a field deployment study) requested a comment field to write a description of what they were seeing. This request is particularly interesting, because none of these participants could think of a way that, as data consumers, they would have a use for this data. They simply "wanted to be able to add a little more data." The disparity between their desires as data collectors and as data consumers reinforces the value in studying both aspects of a citizen science application."

Even more importantly, testing in the field will help to uncover any 'structural' issues such as visibility of the screen in poorly lit areas, taking a photo one-handed if an object must be held simultaneously, or the importance of building data storage into your app for when the participant might be out of reception range—allowing for the uploading of the data when an internet or data connection has been re-established.

Further insights into the UI/UX particularities can be found in the literature with respect to designing virtual citizen science projects [31, 40], how technology is being applied in interesting new ways [2], case studies reporting on mobile application based projects [41–43], and best practice from the field of Human Computer Interaction (HCI) as applied to biodiversity citizen science [29].

## 2.5 Motivational Factors for Participation

There is a great deal in the citizen science literature about the motivations to participate in projects, how to attract participants based on those motivations, and how to maintain their involvement over the longer term [44, 45]. Because participants in citizen science are donating their time and effort freely, project initiators also have a moral obligation and duty to care for their volunteers, and to ensure that the project 'gives back' in keeping with those motivations. A good participant-centred design process will bring the relevant motivations of any given project to the foreground, which are likely to fall into one or more of the following motivational categories:

1. Learning about science [45, 46].
2. Making a contribution to science/collective motivations that are associated with the overall goal of the movement, including a sense of altruism [46, 47, 49].
3. Social proof of seeing that an action is valued and that others have engaged in that action/social motivations that reflect the importance of recognition by others/recognition and attribution [45, 46, 49].
4. Reward-based motivations [49].
5. Intrinsic motivations, where a participant contributes because of personal interest and enjoyment [50].

Both the project flow of tasks and the underlying mobile app need to be designed to take these motivational factors into account, including specific features to support them. This will enhance engagement at the recruitment phase, as well as over the entire length of the project [48].

### 2.5.1 Learning About Science: Supporting Shared Learning

In citizen science projects where the participant is acting independently (such as online, or with a mobile app outside the context of an organiser-led field project) learning takes place at every step, from the initial engagement with an app or platform, to actually doing the task, and beyond. Learning and communication are reciprocal (what is called two-way interaction, as is common in Bioblitz events— see [51]) and occur in tangible, as well as in intangible ways. Kloetzer et al. [52] described various forms of learning and found that most learning occurs in an informal context. Therefore, the value of unstructured learning and communication has to be recognised, with space created for this to take place.

Platforms in which the community helps identifying (and validating) the observations, such as Natusfera[30] (Fig. 2.3), are a good example for how to support unstructured learning, and a powerful tool to engage untrained people who will learn progressively with the help of the community.

The mobile application itself will have limited means to support learning about the object of observation or measurement in the field. However, the project website

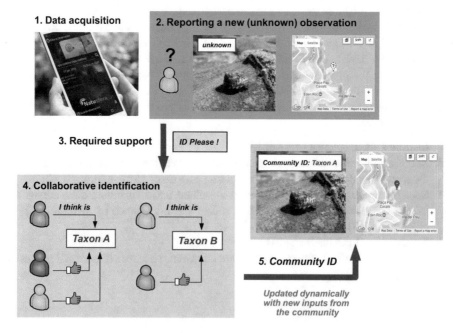

**Fig. 2.3** Schematics of how the community may help in identifying observations that the participants are not able to identify by themselves (the process can also be used also to validate or correct proposed identifications)

---

[30]natusfera.gbif.es.

can fill this gap, with ongoing news updates from the project organisers, shared learning from the researchers, and signposting further information for reading and deeper understanding of the science.

### 2.5.2   Making a Contribution to Science: Designing for Two-Way Communication

News sharing channels by researchers are an important way to feel part of a bigger endeavour, but a real sense of contributing to science can only be achieved by two directions of communication—between practitioners and participants, and among participants themselves. Any citizen science project should plan for and provide two-way communication. As Jennet et al. [53] stated "It is important to provide users with tools to communicate in order to supporting social learning, community building and sharing."

Mobile applications provide a unique opportunity to embed these communications channels within the app itself, such as sending feedback to the researchers via a built-in text messaging function, or sharing notes and observations with the community via a comments function. For example, the EpiCollect[31] app (Fig. 2.4)

**Fig. 2.4** Embedded Google Talk function in EpiCollect mobile app for field data collection (source: journals.plos.org/plosone/article/figure?id=10.1371/journal.pone.0006968.g003)

---

[31] epicollect.net.

for collecting field data via bespoke forms, has embedded the Google Talk instant messenger into the app for instant field communications with the 'curator' of the project. This requires the participant to have a Gmail account, which then automatically stores transcripts for future reference [54].

### 2.5.3  Social Proof: Building and Supporting an Active Community

Social proof can be understood as a psychological factor that comes into play when we see evidence that other people are enjoying an activity, and that we might enjoy it too [46]. Social proof also refers to an activity perceived as worthwhile because there is a community of people already engaged in this activity [49].

Both require visibility of the community, and for the activities of the community to be evident. Koh et al. [55] have identified offline and online interaction as key drivers for community building and collaboration:

> "Leaders of robust, sustainable virtual communities find ways to strengthen their members' sense of social identity and motivate their participation in the community's activities. Understanding virtual community development provides a foundation for facilitating collaboration and learning among individuals separated by physical distance and organisational boundaries."

Social presence in physically dispersed communities can be aided by communication tools such as live text, chat and video interfaces, and also by opportunities to form stronger social bonds in person, at events or group field excursions. Kim [56] suggests four factors for building sustainable communities: clear purpose or vision, clear definition of members' roles, leadership by community moderators, and online/offline events.

Online discussion forums are a simple but effective means of achieving this, as well as community-oriented social media channels such as Facebook Groups. When implementing such community building tools, it is vital that a communication plan with resourcing also be in place, so that participants frequently hear back from the project initiators and researchers.

### 2.5.4  Reward-Based Motivations: Sustaining Long-Term Engagement

The motives of volunteers may be different when participating in open-ended citizen science projects, and can also change over of time. Long-term projects that incorporate little or no user-rewards are likely to hit a plateau in the number of users and encounter challenges in recruiting them [57].

A range of different reward systems may be considered in these cases, which focus on maximizing both the quality and quantity of the data collected, as well as on retention of volunteers. Within citizen science projects, these can be divided into two main types: symbolic and non-symbolic [2, 58], examples of which are shown in Appendix D.

An example of different reward systems implemented in two recent campaigns run by the Geo-Wiki[32] team [59, 60] is Picture Pile. It is a cross-platform application that is designed as a generic and flexible tool for ingesting satellite imagery for rapid classification. The application involves simple micro-tasks, where the user is presented with satellite images and is asked a straight-forward yes/no question. Using this app, campaigns have been run with both symbolic rewards and no rewards:

- **Non-symbolic rewards:** In one campaign, volunteers were asked to identify the presence or absence of cropland from very high resolution satellite images and geotagged photographs. Each week, the top three players with the highest score were added to a list of weekly winners. The campaign ran for around 6 months, after which three people from the list of weekly winners were randomly drawn to win prizes, which included an e-reader, a smartphone and a tablet.
- **Symbolic rewards or no rewards:** In a second campaign, volunteers were asked to look at pairs of very high resolution satellite images "before" and "after" Hurricane Matthew hit Haiti to identify the presence of any visible building damage. There were no rewards although personal performance ratings and ratings on a leaderboard were provided to incentivize participation.

Both campaigns were successful in terms of the data collected, despite the different reward systems used. The difference was in the type of task undertaken by the volunteers, which attracted individuals with different underlying motivations.

## 2.6 Discussion and Conclusions

Apps and platforms used through mobile devices enable citizens to provide timely geospatial information that contributes to scientific understanding and decision-making for environmental and biodiversity citizen science. In this chapter, we encourage initiators of new mobile-based citizen science projects to (1) follow existing data and web standards where possible, (2) collaborate and consult with the target audience of participants early and often, (3) not reinvent the wheel, (4) build

---

[32]geo-wiki.org.

on existing UI/UX expertise regarding the development of mobile applications, and (5) factor in motivational considerations throughout.

Using accepted and well-established open standards helps to ensure reliability and interoperability with other tools. The number and range of standards is indeed vast, as only partially illustrated by Appendix A, yet knowing this will provide a solid base for taking an informed decision within each specific project.

Useful guidelines, best practices and other training material to assist in the choice of standards are being worked on in the form of publications [21, 22, 61], in the context of the CSA International Working Group on Data and Metadata,[33] or OGC's Citizen Science Domain Working Group.[34]

Before embarking on the process of building an app and its associated website, effort needs to be made to not reinvent the wheel by looking for open source code repositories and the re-usable elements of other projects. The early years of citizen science apps have seen considerable (near) reproduction of already existing apps (e.g. for noise monitoring apps shown in Appendix E), yet sufficient mature applications now exist, as shown in Appendix C, for reuse and re-usability to become the norm.

However, it can be a challenge to find existing apps with development documentation that is thorough, up-to-date, and also includes feedback (such as reporting test results' user experiences). A discussion of how it can sometimes be more time consuming to re-use a ready-to-use tool than to build a new one can be found in the Schade et al. [18] assessment of invasive alien species apps and their potential for reuse. The need for a 'neutral' cross-topic inventory to aid the discovery and reuse of existing apps is clear. This is also exemplified in Appendix E, which provides a snapshot of the large number of noise pollution apps that have been re-created each time from scratch.

Of even more importance than the ease of re-using existing apps, source code, platforms and standards, is the fact that this can significantly lower the investment cost in terms of money, effort and expertise. It takes a professional software development approach to develop sufficiently mature applications, which is often neither in the scope of scientific projects, nor accounted for in the budget planning.

There are naturally tensions between different points of view, even inside the community of citizen science practitioners, with some advocating for a smaller number of platforms and systems in the name of efficiency and economies of scale; whereas others point out the need for innovation and new approaches. In this chapter we hope to have highlighted the wide range of choice available to project designers to meet the unique needs of their project and local context.

In conclusion, we highlight the general principles of citizen science, as they are expressed in the ECSA Ten Principles of Citizen Science[35] as a guiding force towards best practice when designers and developers are embarking on a new citizen

---

[33]citizenscience.org/2015/11/12/introducing-the-data-and-metadata-working-group.

[34]opengeospatial.org/projects/groups/citizenscience.

[35]ecsa.citizen-science.net/sites/default/files/ecsa_ten_principles_of_ citizen_science.pdf.

science project. Undoubtedly, technology is only part of the story. New technologies open up many new possibilities, including the capacity to scale globally, yet a local focus and community-mindedness will always be needed.

# References

1. Jennett, C, Furniss, D J, Iacovides, I, Wiseman, S, Gould, S J J and Cox, A L "Exploring Citizen Psych-Science and the Motivations of Errordiary Volunteers", Human Computation 1 (2), 200–218. (2014)
2. Newman G, Wiggins A, Crall A, Graham E, Newman S and Crowston K, The future of citizen science: Emerging technologies and shifting paradigms, Frontiers in Ecology and the Environment 10(6): 298–304, (2012) https://doi.org/10.1890/110294
3. Raddick MJ, Bracey G, Gay PL, Lintott CJ, Cardamone C, Murray P, Galaxy Zoo: Motivations of citizen scientists, Astronomy Education Review, 12(1) (2013)
4. Wiggins, A, and Crowston, K, From conservation to crowdsourcing: a typology of citizen science, In Proc. of 44th Hawaii International Conference on System Sciences (HICSS '10) (2011)
5. Bonney, R, Shirk, JL, Phillips, TB, Wiggins, A, Ballard, HL, Miller-Rushing, AJ and Parrish, JK, Next Steps for Citizen Science, Science 343 (6178), 1436–1437 (2014)
6. Haklay M, Citizen Science and Volunteered Geographic Information: Overview and Typology of Participation. In: Crowdsourcing Geographic Knowledge, Edited by D Z Sui, S Elwood and M F Goodchild, Dordrecht, Netherlands: Springer, 105–122 (2013)
7. Teacher AGF, Griffiths DJ, Hodgson DJ, Inger R, Smartphones in ecology and evolution: a guide for the app-rehensive, Ecology and Evolution 3(16): 5268–5278 (2013)
8. Pettibone L, Vohland K, Ziegler D, Understanding the (inter)disciplinary and institutional diversity of citizen science: A survey of current practice in Germany and Austria. PLoS ONE12(6): e0178778, (2017) https://doi.org/10.1371/journal.pone.0178778
9. Pocock MJO, Tweddle JC, Savage J, Robinson LD, Roy HE, The diversity and evolution of ecological and environmental citizen science, PLoS ONE 12(4): e0172579 (2017) https://doi.org/10.1371/journal.pone.0172579
10. Chandler M, See L, Copas K, Bonde AMZ, Claramunt B, Danielsen F, Legind JK, Masinde S, Miller-Rushing AJ, Newman G, Rosemartin A, Turak E, Contribution of citizen science towards international biodiversity monitoring, Biologica Conservation (2016) http://dx.doi.org/10.1016/j.biocon.2016.09.004
11. Sheth AP, Citizen Sensing, Social Signals, and Enriching Human Experience, IEEE Internet Computing, 13(4), 87–92 (2009) http://corescholar.libraries.wright.edu/knoesis/728
12. Ceccaroni L, Bowser A and Brenton P, Civic Education and Citizen Science: Definitions, Categories, Knowledge Representation, Analyzing the Role of Citizen Science in Modern Research, IGI Global, 1–23 (2017)
13. Eitzel MV, Cappadonna JL, Santos-Lang C, Duerr RE, Virapongse A, West SE, Conrad C, Kyba M, Bowser A, Cooper CB, Sforzi A, Metcalfe AN, Harris ES, Thiel M, Haklay M, Ponciano L, Roche J, Ceccaroni L, Shilling FM, Dörler D, Heigl F, Kiessling T, Davis BY, Jiang Q, Citizen Science Terminology Matters: Exploring Key Terms. Citizen Science: Theory and Practice. 2(1), p.1. (2017) http://doi.org/10.5334/cstp.96
14. Ceccaroni L and Piera J, Analyzing the Role of Citizen Science in Modern Research, IGI Global, 25 Oct (2016)
15. Bizer C, Heath T and Berners-Lee T, Linked Data – The Story So Far, International Journal on Semantic Web and Information Systems 5(3), pp1–22 (2009).
16. Groom Q, Weatherdon L and Geijzendorffer IR, Is citizen science an open science in the case of biodiversity observations?, Journal of Applied Ecology, 1–6 (2016) doi: 10.1111/1365-2664.12767

17. Schade S and Tsinaraki C, Survey report: data management in Citizen Science projects, JRC Technical Report JRC101077 (2016) DOI: 10.2788/539115
18. Schade S, Tsinaraki C and Roglia E, Scientific Data from and for the Citizen, First Monday, August 2017, Volume 22, Number 8 (2017) DOI: http://dx.doi.org/10.5210/fm.v22i8.7842
19. Scassa T and Haewon C, Managing Intellectual Property Rights in Citizen Science: A Guide for Researchers and Citizen Scientists, Washington, DC: Woodrow Wilson International Center for Scholars (2015) http://www.wilsoncenter.org/publication-series/commons-lab
20. Bowser A, Shilton K, Preece J, and Warrick E, Accounting for Privacy in Citizen Science: Ethical Research in a Context of Openness, In Proceedings of the 2017 ACM Conference on Computer Supported Cooperative Work and Social Computing (CSCW '17), ACM, New York, NY, USA, 2124–2136 (2017) DOI: https://doi.org/10.1145/2998181.2998305
21. Bastin, L, Sven S, and Schill C, Data and metadata management for better VGI reusability. Citizen Sensor (2017): 249.
22. Williams J, et al. Citizen-science data, how should you maximise their impact and sustainability? In: Hecker, S., Haklay, M., Bowser, A., Makuch, Z., Vogel, J., & Bonn, A. (2018). Citizen science-Innovation in open science, society and policy.
23. Schnetzer J, Kopf A, Bietz MJ, Buttigieg PL, Fernandez-Guerra A, Ristov AP, Kottmann R, MyOSD 2014: Evaluating Oceanographic Measurements Contributed by Citizen Scientists in Support of Ocean Sampling Day, Journal of Microbiology & Biology Education, 17(1), 163–171 (2016) http://doi.org/10.1128/jmbe.v17i1.1001
24. Bonney R, Cooper CB, Dickinson J, Kelling S, Phillips T, Rosenberg KV and Shirk J, Citizen science: A developing tool for expanding science knowledge and scientific literacy, Bioscience, 59, 11, 977–984 (2009)
25. Rotman D, Collaborative science across the globe: The influence of motivation and culture on volunteers in the United States, India and Costa Rica, Ph.D. Dissertation, University of Maryland (2013) http://drum.lib.umd.edu/handle/1903/14163
26. Curtis V, Online citizen science projects: an exploration of motivation, contribution and participation, Ph.D. thesis, The Open University (2015)
27. Crowston K and Prestopnik NR, Motivation and Data Quality in a Citizen Science Game: A Design Science Evaluation, In: Proceedings of HICSS 2013. IEEE, pp. 450–459 (2013)
28. Rotman D, Hammock J, Preece J, Hansen D, Boston C, Bowser A and He Y, Motivations Affecting Initial and Long-Term Participation in Citizen Science Projects in Three Countries, In: Proceedings of iConference 2014, iSchools (2014)
29. Preece J, Citizen Science: New Research Challenges in HCI, International Journal of Human-Computer Interaction 32, 8, 585–612 (2016) http://www.tandfonline.com/doi/full/10.1080/10447318.2016.119415
30. Eveleigh AMM, Jennett C, Blandford A, Brohan P and Cox AL, Designing for dabblers and deterring drop-outs in citizen science. In: Proceedings of the IGCHI Conference on Human Factors in Computing Systems (CHI'14), New York, NY, U.S.A.: ACM Press, 2985–2994 (2014) DOI:10.1145/2556288.2557262
31. Jennett C and Cox A, Eight Guidelines for Designing Virtual Citizen Science Projects, Citizen + X: Volunteer-Based Crowdsourcing in Science, Public Health, and Government: Papers from the 2014 HCOMP Workshop (2014)
32. Fielding RT, Chapter 5: Representational State Transfer (REST), Architectural Styles and the Design of Network-based Software Architectures (Ph.D.). University of California, Irvine (2000)
33. Anderson J, McRee J, Wilson R, Effective UI: The Art of Building Great User Experience in Software, O'Reilly Media (2010)
34. Wroblewski L, Mobile First, A Book Apart (2011)
35. Ries E, The Lean Startup: How Today's Entrepreneurs Use Continuous Innovation to Create Radically Successful Businesses, Crown Business, New York (2011)
36. Bieller E, How To Design A Mobile App Using User Interface Design Principles, 06 September, Career Foundry Blog (2016) https://careerfoundry.com/en/blog/ui-design/how-to-design-a-mobile-app-using-user-interface-design-principles/

37. Zeldman J, Taking Your Talent to the Web: A Guide for the Transitioning Designer, New Riders Publishing (2001)
38. Edwards R, Phillips TB, Bonney R and Mathieson K, Citizen Science and Science Capital, Stirling: University of Stirling (2015)
39. Conrad CC and Hilchey KG, A review of citizen science and community-based environmental monitoring: issues and opportunities, Environmental Monitoring and Assessment 176: 273 (2011) https://doi.org/10.1007/s10661-010-1582-5
40. Yadav P and Darlington J, Design Guidelines for the User-Centred Collaborative Citizen Science Platforms. Human Computation 3:1:205–211 (2016)
41. Maisonneuve N, Stevens M, Niessen ME, Steels L, NoiseTube: Measuring and mapping noise pollution with mobile phones, In: Athanasiadis IN, Rizzoli AE, Mitkas PA, Gómez JM (eds) Information Technologies in Environmental Engineering, Environmental Science and Engineering, Springer, Berlin, Heidelberg (2009)
42. Kim S, Robson C, Zimmerman T, Pierce J and Haber E, Creek Watch: Pairing Usefulness and Usability for Successful Citizen Science, CHI 2011, May 7–12, Vancouver, BC, Canada (2011)
43. Traynor B, Lee T, Duke D, Case Study: Building UX Design into Citizen Science Applications, In: Marcus A, Wang W (eds) Design, User Experience, and Usability: Understanding Users and Contexts, DUXU 2017, Lecture Notes in Computer Science, vol 10290 Springer, Cham (2017)
44. Mueller M, Tippins D and Bryan L, The future of citizen science. Democracy & Education 20(1): 1–12 (2012)
45. Rotman D, Preece J, Hammock J, Procita K, Hansen D, Parr C, Lewis D and Jacobs D, Dynamic changes in motivation in collaborative citizen-science projects, In: Proceedings of CSCW 2012, ACM Press, 217–226 (2012)
46. Lee, Miller and Crowston, Recruiting Messages Matter: Message Strategies to Attract Citizen Scientists, CSCW '17 Companion, February 25 - March 01, 2017, Portland, OR, USA (2017) http://dx.doi.org/10.1145/3022198.3026335
47. Land-Zandstra A, Devilee J, Snik F, Buurmeijer F, and van den Broek J, Citizen science on a smartphone: Participants' motivations and learning, Public Understanding of Science, Vol 25, Issue 1, 45–60 (2015)
48. Nov O, Ofer A and David A, Technology-Mediated Citizen Science Participation: A Motivational Model, ICWSM (2011)
49. Klandermans B, Collective political action, Oxford handbook of political psychology: 670–709 (2003)
50. Geoghegan H, Dyke A, Pateman R, West S and Everett G, Understanding motivations for citizen science, Final report on behalf of UKEOF, University of Reading, Stockholm Environment Institute (University of York) and University of the West of England (2016)
51. Jennett C, Cognetti E, Summerfield J and Haklay M, Usability and interaction dimensions of participatory noise and ecological monitoring, In Participatory Sensing, Opinions and Collective Awareness, 201–212, Springer International Publishing (2017)
52. Kloetzer L, Schneider D, Jennett C, Iacovides I, Eveleigh A, Cox AL, Gold M, Learning by volunteer computing, thinking and gaming: What and how are volunteers learning by participating in Virtual Citizen Science? ESREA 2013, Germany (2013)
53. Jennett C, Kloetzer L, Schneider D, Iacovides I, Cox AL, Gold M, Fuchs B, Eveleigh A, Mathieu K, Ajani Z and Talsi Y, Motivations, learning and creativity in online citizen science, Journal of Science Communication 15 (3) (2016)
54. Aanensen DM, Huntley DM, Feil EJ, and Spratt BG, EpiCollect: linking smartphones to web applications for epidemiology, ecology and community data collection, PloS one, 4(9), e6968 (2009)
55. Koh J, Kim A, Butler B and Bock G, Encouraging Participation in Virtual Communities, Communications of the ACM, Vol.50. No. 2 (2007)
56. Kim A, Community Building on the Web, Peachpit Press, Berkeley, CA (2000)
57. Sullivan BL, Wood CL, Iliff MJ, Bonney RE, Fink D and Kelling S, eBird: A citizen-based bird observation network in the biological sciences, Biological Conservation, Elsevier BV (2009) https://doi.org/10.1016/j.biocon.2009.05.006

58. Wiggins A, Crowdsourcing Scientific Work: A Comparative Study of Technologies, Processes, and Outcomes in Citizen Science, The School of Information Studies- Dissertations, Paper 72 (2012)
59. Danylo O, Sturn T, Giovando C, Moorthy I, Fritz S, See L, Kapur R, Girardot B, Ajmar A, Giulio Tonolo F, Reinicke T, Mathieu P and Duerauer M, Picture Pile: A citizen-powered tool for rapid post-disaster damage assessments, Geophysical Research Abstracts, Vol. 19, EGU2017-19266, 2017 EGU General Assembly 2017 (2017)
60. Sturn T, Wimmer M, Salk C, Perger C, See L, and Fritz S, Cropland Capture — A Game for Improving Global Cropland Maps, In: Foundatoins of Digital Games, 22–25 June 2015, California (2015)
61. Bastin L, Schade S and Mooney P, Standards, encodings and tools for assessing fitness for purpose, In: Bordogna, Gloria, and Paola Carrara, eds. Mobile Information Systems Leveraging Volunteered Geographic Information for Earth Observation. Vol. 4. Springer, 2017.
62. Radicchi, A. The HUSH CITY app. A new mobile application to crowdsource and assess "everyday quiet areas" in cities. Invisible Places. Sound, Urbanism and the Sense of Place, 511-528.

# Chapter 3
# A Toolbox for Understanding and Implementing a Citizens' Observatory on Air Monitoring

Hai-Ying Liu, Mike Kobernus, Mirjam Fredriksen, Yaela Golumbic, and Johanna Robinson

**Abstract** This chapter explores the growing trend of using innovative tools, particularly low cost micro-sensors and mobile apps, to facilitate the citizen participation process within an environmental monitoring programme. Special focus will be put on tools that are linked to major initiatives that form citizens' observatories (CitObs) on air quality and that address novel ways to involve citizens. On the basis of providing an overview of a Citizens' Observatory' Toolbox (COT) developed in the EU FP7 CITI-SENSE project, this chapter introduces a number of tools that together provide the foundation of a complete citizens' observatory. We present these tools from two perspectives, the users' and the developers' perspective. Special emphasis will be placed on those tools developed specifically for the nine case study locations within the CITI-SENSE project and on the technical elements and frameworks that can be further exploited for the creation of new citizens' observatories in the future. In addition, this chapter highlights the usage of these tools in order to support citizens' involvement within environmental monitoring of air quality, particularly for the collection of data and observations via sensors, mobile apps and surveys. This chapter will also touch on analysis and visualisation of observations through software widgets and web portals.

H.-Y. Liu (✉) · M. Kobernus · M. Fredriksen
NILU – Norwegian Institute for Air Research, Kjeller, Norway
e-mail: hai-ying.liu@nilu.no

Y. Golumbic
TECHNION – Israel Institute of Technology, Technion City, Haifa, Israel

J. Robinson
JSI – Jožef Stefan Institute, Ljubljana, Slovenia

© Springer International Publishing AG, part of Springer Nature 2018
A. Joly et al. (eds.), *Multimedia Tools and Applications for Environmental & Biodiversity Informatics*, Multimedia Systems and Applications,
https://doi.org/10.1007/978-3-319-76445-0_3

## 3.1  Introduction

Within environmental monitoring there is a paradox. Although this is undisputedly the age of big data, for many organisations engaged in collecting large amounts of environmental data there are still challenges due to a lack of relevant data, information and knowledge [20]. At the same time, the availability of relatively cheap, Internet-connected, programmable, sensor-laden smart-phones and the explosive growth of communication devices has vastly increased the potential for personal data-collection applications. As a result, there has been a boom within Citizen Science (CS) and Citizens' Observatories (CitObs) related projects [14, 16] which enable members of the public to play an active part in environmental monitoring by exploiting these innovative technologies across the globe [7, 13]).

With the advent of new low-cost sensor technologies, monitoring air pollution can now be performed by any interested individual [8, 15]. Novel sensor technologies provide opportunities to monitor air quality at spatial resolutions not possible to achieve with the highly expensive monitoring systems used by research organisations [14, 18]. Low-cost personal sensors are small, portable and easy to use and have created a paradigm shift where citizens are enabled to directly monitor their environment. In a very real sense, citizens can now contribute to monitoring their environment in a manner that is useful both from a perspective of awareness raising, as well as providing usable data for research purposes, despite the negative side of the low-cost sensors and their usage (e.g., sensitivity, stability, data quality, etc.) [2].

Within this context, several projects have explored the possibility of using sensors and/or mobile apps for air quality monitoring or estimation. For example, the following projects all use microsensors for air quality measurement: CITI-SENSE,[1] hackAIR,[2] CAPTOR[3] and AirTick.[4] In particular, CITI-SENSE developed and demonstrated a sensor-based citizens' observatory community for urban air quality monitoring within nine European cities, while hackAIR aims to complement official particulate matter (PM) data with community-driven data sources, for collecting, analysing and sharing air quality measurements to community members through low-cost open hardware sensors assembled by citizens, web and/or mobile phones. CAPTOR addresses the general air pollution problem, which is depicted by monitoring tropospheric ozone ($O_3$) and by engaging a network of local communities to use low-cost sensors for data collection. AirTick uses crowdsourced photos combined with official air quality data and machine-learning to develop a method for predicting pollution levels based on the images alone. The commonality of these projects are their usage of tools such as micro-sensors and/or mobile apps

---

[1]http://co.citi-sense.eu.

[2]www.hackair.eu.

[3]www.captor-project.eu.

[4]http://crowdsourced-transport.com/airtick-air-quality-monitoring-from-selfies.

to enable the general public to contribute to air quality monitoring and to improve their environmental awareness and initiate behavioural changes. The differences between the projects are the focus of the air pollutants and the pilot testing scale, e.g., CITI-SENSE focused on gases NO, $NO_2$, $NO_x$, $O_3$, CO, $SO_2$, particulates $PM_1$, $PM_{2.5}$, $PM_{10}$ and TPC (Total Particle Count) and %RH (Relative humidity), pod temperature, atmospheric pressure and noise for nine city locations (Barcelona, Belgrade, Edinburgh, Haifa, Ljubljana, Oslo, Ostrava, Vienna, Vitoria), hackAIR measured $PM_{2.5}$ and $PM_{10}$ in two European countries (Germany and Norway), CAPTOR monitored $O_3$ in three European regions (Catalonia, Spain; Po Valley, Italy; Burgenland, Steiermark and Niederösterreich, Austria). There are also different approaches for air quality estimation by using images, e.g., hackAIR retrieves aerosol optical depth (AOD) values from user generated images using calculated colour R/G (Red/Green) or G/B (Green/Blue) values from images, estimating the current particulate pollution. AirTick extracts the haziness component from outdoor images, and passes this component to the Deep Neural network Air quality estimator (DNA), to produce an estimated Pollutant Standards Index (PSI) value for the user.

In this chapter, we present a Citizens' Observatory Toolbox (COT) with a number of applications and tools that together provide the foundation of a complete citizens' observatory. The COT was developed within the EU FP7 CITI-SENSE project for collecting citizen-contributed observations of air quality in cities [4, 14, 18]. We present these tools from various perspectives; in particular, the users' and the developers' perspectives and include real-world examples of their use. Special emphasis is placed on the technical elements and frameworks of the COT that can be reused for the creation of new CS and CitObs by future projects or initiatives.

## 3.2  Overview of the Citizens' Observatory Toolbox

The main goal of the COT is to provide access to a number of custom made tools with guidelines, that can be used now and in the future by citizens, scientists and interest groups, as well as any commercial interest. The COT includes any resources, procedures, software, hardware or services that can be used to support citizens in participating within environmental monitoring and enable them to contribute to community based environmental decision making. The COT consists of seven areas: Methods, Data, Web portals, Widgets, Sensors, Surveys and Mobile apps, which are defined in the COT 'flower' (Fig. 3.1, [12]). However, while these various aspects of the COT are segregated by type, they fall within one or more of the following topics:

- Server side management of **Sensors** data and other types of Surveys data: **Surveys** provide a good approach for interaction with citizens; Sensors are one of the main tools to engage citizens.
- User applicability, using **Mobile Apps** and **Web Portals**: Web Portals represents the most visible interface to the CITI-SENSE result and data. Mobile Apps are important to support citizen's participation.

**Fig. 3.1** Citizens'
observatory toolbox

- **Methods** for generating data, access/download data and view/visualise data with references to the relevant documents; and
- **Data:** raw data and post-processed data.
- **Widgets** and code-snippets: Widgets are reusable user interface components that can be reused in future CitObs.

## 3.3   Citizens' Observatory Toolbox – Users' and Developers' Perspective

The COT contains a number of elements that together provide the foundation of a complete CitObs, with tools for data collection, methodology for how and what to collect, as well as dissemination platforms via web portals and phone apps to display and distribute information. In all, the COT is a comprehensive collection of useful mechanisms for the collection and distribution of citizen gathered information.

But while the COT is designed to be easy to use with no steep learning curve, it must also provide useful information in a manner that is easily accessible and understandable to various users who will use the citizen data and information. That is, citizens and scientists. To that end, the system developers and business analysts employed within the project had to balance the conflicting needs of both groups. Careful planning ensured that the system was designed in an optimal manner, with little to no refactoring necessary in the later stages of the project.

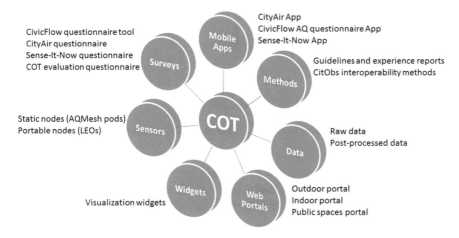

**Fig. 3.2**  Citizens' observatory toolbox overview: users' perspective

In the following, we will present the COT from the users' and developers' perspective. From the users' perspective, we ensure that the COT answers questions such as "what is this tool and how can it be used?". From the developers' perspective, we describe the COT in a manner that addresses questions like "how was this tool developed and how may it be further used?".

## 3.3.1   Users' Perspective

The elements provided through the users' perspective of the COT, with an emphasis on the elements that have been developed for the nine city locations, include support manuals and user guides for the collection of objective data and subjective or personal observations through **Sensors, Mobile apps** and **Surveys**, and also elements and methodologies for analysis and visualisation of observations via **Widgets** and **Web portals**. A simple definition of objective observations are those that can be described as quantitative, and may be made with sensors, Fig. 3.2 is an overview of the COT contents from the users' perspective.

- **Sensors**: A set of environmental micro-sensor platforms were developed for both static and portable air quality sensors, including monitoring of NO, $NO_2$, $O_3$, $CO/CO_2$, $PM_{2.5}$, $PM_{10}$, Radon, temperature and humidity [5]. Static nodes (AQMesh pods) and portable nodes (LEOs-Little Environmental Observatory) can be deployed with different stakeholders (Fig. 3.3). The portable nodes are coupled to the ExpoApp smartphone application (Android), which enable the connection between the LEOs and the smartphone. The ExpoApp also collects geolocation and accelerometer data whenever and wherever the LEOs is carried. Both the AQMesh pods and the LEOs automatically and wirelessly transmit

**Fig. 3.3** Example of the sensors units (left: LEOs unit; middle: AQMesh pod; right: Radon unit)

encrypted data to dedicated Spatial and Environmental Data Service (SEDS) and Web Feature Service (WFS) databases, where they can be further processed for visualisation and evaluation for usefulness by users.

- **Mobile Apps:** A set of mobile apps to support user perceptions and users answers to surveys and questionnaires, with a focus on air quality questions. The CityAir app (Android and iOS) can be used to collect and geo-locate personal perceptions on the level and source of air pollution emissions in real-time. The diagram in Fig. 3.4 [6] shows the users' interaction with the CityAir App,[5] and subsequent extensions to other channels. The **CivicFlow AQ questionnaire**[6] can be applied to study public general knowledge of air quality issues and associated information services, and perception in cities. The **Sense-It-Now App**[7] (Android) reads and displays data from different sensors. It also gives users the possibility to add perceptions about their current environment.[8]
- **Surveys:** A set of web-based and smartphone based surveys to study public knowledge and their level of awareness of air quality related issues,[9] and their evaluation about effectiveness and usefulness of the CitObs COT.
- **Widgets:** A set of reusable widgets to visualise the collected observations in different ways, typically time and location-based queries, that can be used both through development in smartphone apps as well as in portal development.[10]
- **Methods:** A set of methods include guidelines for how to use sensors, mobile apps and portals in order to support citizens' involvement and empowerment in environmental monitoring of air quality.[11]
- **Web Portals:** A set of web portals for information, dissemination and communication.[12] It includes the CitObs central web portal, and its thematic areas (i.e.,

---

[5]https://play.google.com/store/apps/details?id=io.cordova.CityAir&hl=en.

[6]http://www.civicflow.com.

[7]https://git.nilu.no/citi-sense/sense-it-now/blame/ada3e0b10fba65b35f673c3b93ff48f7a3bd1ade/CSToolboxTest/PostRequest.html.

[8]http://co.citi-sense.eu/CitizensObservatoriesToolbox/MobileApps/SENSE-IT-NOW.aspx.

[9]http://co.citi-sense.eu/CitizensObservatoriesToolbox/Surveys.aspx.

[10]http://co.citi-sense.eu/CitizensObservatoriesToolbox/Widgets.aspx.

[11]http://co.citi-sense.eu/CitizensObservatoriesToolbox/Methods.aspx.

[12]http://co.citi-sense.eu/CitizensObservatoriesToolbox/WebPortals.aspx.

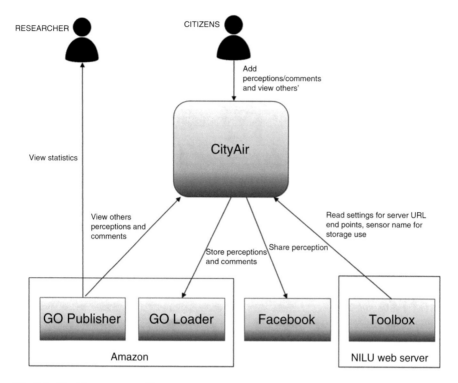

**Fig. 3.4** CityAir app context diagram

outdoor air quality in cities, indoor air quality in schools and environmental quality in public spaces), and individual web portals of participating cities (Fig. 3.5 [11]).

## 3.3.2  Developers' Perspective

From developers' perspective, we focus on how the software and tools from the COT were developed and might be used in future CitObs projects (Fig. 3.6). There was considerable emphasis placed on creating "generic enablers", that is, elements that could be reused, like building blocks, for the creation of new CitObs. Figure 3.6 depicts the elements provided through the developers' perspective of the COT, including:

- **Methods:** How Citizens' Observatory data is collected, managed and used?
- **Data:** How the collected data are transformed, stored and accessed?
- **Web Portals:** How data can be made available through web pages and the concept of CitObs?
- **Widgets:** How data is being visualised?

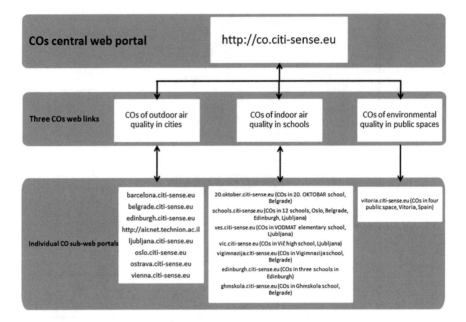

**Fig. 3.5** Triple layers of citizens' observatories web portals (COs = CitObs)

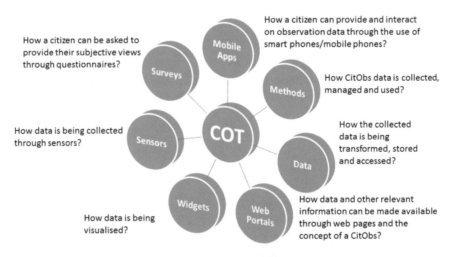

**Fig. 3.6** Citizens' observatory toolbox overview: developers' perspective

- **Sensors:** How data is being collected through sensors?
- **Surveys:** How a citizen can be asked to provide their subjective views (e.g., perception of air quality in their surroundings) through questionnaires?
- **Mobile Apps:** How a citizen can provide and interact on observation data through the use of smart phones/mobile phones?

### 3.3.2.1   Architecture

To realize these elements, the **Architecture** was of prime importance, as the solution needed to meet the immediate needs of the project, yet still be flexible enough to accommodate undefined requirements in the future [9]. Therefore, the project employed a scalable and flexible architecture with an option for multiple front end sensor systems as well as multiple mobile apps for data collection and visualization. A generic framework for multiplatform mobile observation apps was provided. The CITI-SENSE Data Model version 2.2 was implemented in a WFS server, and supports the storage of air quality observations as well as subjective data, such as questionnaires. The WFS server supports access to the data through multiple presentation formats such as XML, CSV and JSON. In addition there is an experimental export to linked data through a SPARQL end-point which enables users to query a database via the SPARQL language. Figure 3.7 shows the technical data flow from the various sensor provider inputs to the SEDS database server with data ingest services and data publication services [1]. Related to the technical interoperability of CitObs in general, the focus here is on supporting heterogeneity among both sensor platforms and for the various user interaction services. The model management interoperability is supported through the use of a common data model in the server.

Figure 3.8 shows the variety of different sensor types and apps on the left and how the data is mapped into a common model in the spatial environmental data server, before being served out again in different forms through application programming interfaces (APIs) (e.g., WFS) or through various representation forms by CSV, XML, JSON, RDF or others [1]. A brief description of the different data sources used is as follows: (1) Ateknea LEOs outdoor portable air monitor and accelerometer: The LEOs are portable sensor packs. It measures NO, $NO_2$ and $O_3$ using electrochemical sensors. The personal sensors together with the ExpoApp smartphone application are developed from Ateknea Solutions.[13] (2) Geotech/AQMesh outdoor static air monitor: Geotech/AQMesh is the commercially available and proven low-cost system for monitoring air quality. The product combines a hardware platform with the latest sensor technology and GPRS (General Packet Radio Service) communication, cloud-based data processing and secure online access. It measures gases NO, $NO_2$, $NO_x$, $O_3$, CO, $SO_2$, particulates $PM_1$, $PM_{2.5}$, $PM_{10}$ and TPC and %RH, pod temperature, atmospheric pressure and noise.[14] (3) Atmospheric Indoor Static Air monitor combines Alphasense' s low-cost and high sensitivity sensors linked with electronics, GPS (The Global Positioning System), GSM (Global System for Mobile communication) and advanced data algorithms. It measures temperature, %RH, $CO_2$, $NO_2$, Particles ($PM_{10}$, $PM_{2.5}$), $O_3$, CO.[15] (4) The Obeo Radon sensor is to remotely monitor radon and/or $CO_2$ levels in indoor areas and buildings. The MMA (Mobile Marketing Association) utilises the GSM/GPRS cellular network to relay the sensor data to a central server available

---

[13]http://ateknea.com.

[14]http://www.aqmesh.com.

[15]http://atmosphericsensors.com/news/model-510-remote-air-quality-monitor.

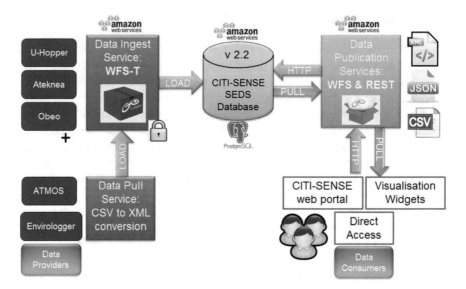

**Fig. 3.7** Citizens' observatories technical interoperability using WFS-T (a transactional web feature service) and REST (representational state transfer) interfaces

from any PC/Mac/Ipad.[16] (5) Tecnalia Kestrel, Smartphone Nexus and Microphone thermal and acoustic monitors consists of Kestrel 4000 sensor device to monitor temperature, %RH, Smartphone Nexus and Microphone to monitor thermal and acoustic level, CityNoise smartphone android app, SENSE-IT-NOW smartphone android app, and a web portal with results for thermal comfort and noise map.[17] (6) U-hopper CivicFlow questionnaire facilitate users to create civic campaigns, trigger citizens participation and get results and analytics.[18] (7) NILU CITI-SENSE user perception air quality smartphone app: CityAir App developed by NILU allows users to rate the air quality in their surroundings and to indicate the source of the pollution and leave a comment.[19]

### 3.3.2.2 Guidelines

Attention was also given to the creation of user guidelines for how to set up and test sensors packages and sensor platforms, open source software for the creation of mobile apps possibly with attached mobile sensors, software for the creation of surveys/questionnaires, a data model and a data storage service for the storage of sensor and human observations and data fusion methodology to add value to the observations of pollutants from low-cost micro-sensors. Currently, these

---

[16]http://co.citi-sense.eu/Portals/1/Templates/Sensor%20guideline/obeo-info-Radon.pdf.

[17]vitoria.citi-sense.eu.

[18]civicflow.com.

[19]https://play.google.com/store/apps/details?id=io.cordova.CityAir.

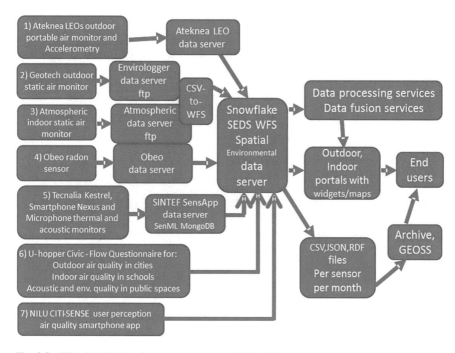

**Fig. 3.8** CITI-SENSE data flow to and from the SEDS WFS server

guidelines are accessible on CitObs central web portal.[20] Figure 3.9 illustrates an example of guideline on how the collected data is connected to the different WFS services created for the communication with the various actors involved. The Data Storage component in the SEDS Platform is implemented as a relational SQL (Structured Query Language) database. The free and open source database technology PostgreSQL is implemented to realise the SEDS Data Repository. PostgreSQL is cross-platform and supports many different operating systems. For storing geographical data, the PostgreSQL database is extended with the PostGIS add-on. PostGIS is a free and open source software application which adds support for geographic objects to the PostgreSQL database. The solution adopted to create the PostgreSQL/PostGIS Relational Database has been the one provided by the Amazon Web Service [1].

Figure 3.10 shows an example of the data fusion process for real-world observations carried out within the CITI-SENSE AQMesh monitoring network deployed in Oslo, Norway. The left panel of the figure illustrates the two input datasets that are required by the data fusion algorithm: the background map shows the long-term average concentration of $NO_2$ as modelled by the EPISODE chemical dispersion model [19], whereas the point makers indicate both the location of the crowdsourced measurement devices as well as the magnitude of the $NO_2$ concentrations observed by each device. It can be seen that in this instance the observations are overall significantly higher than the long-term average concentrations [18]. When the data

---

[20]http://co.citi-sense.eu/CitizensObservatoriesToolbox/Methods.aspx.

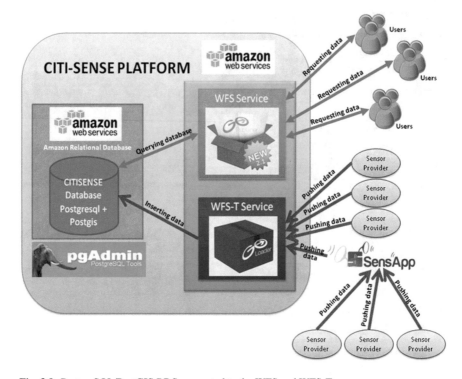

**Fig. 3.9** PostgreSQL/PostGIS RDS connected to the WFS and WFS-T

fusion algorithm is applied to these databases the resulting concentration field (right panel) is much more consistent with the observations [18]. Each fused concentration field is associated with a map of uncertainty (bottom left panel) which illustrates qualitatively how the reliability of the mapping results varied in space and gives quantitative information about each grid cell's mapping uncertainty [18]. Finally, the bottom right panel of Fig. 3.10 shows the basemap correction, i.e., the amount by which each grid cell of the basemap (top left panel) had to be adjusted in order to achieve the data fusion result (top right panel). In this case all correction values are positive as all of the crowdsourced observations were significantly higher than concentrations given by the basemap, however the correction can also take negative values [18].

The results indicate that the using data assimilation and data fusion methods has significant potential for generating realistic and accurate maps of urban air quality in an up-to-date fashion, particularly given the likely future evolution of the sensor devices [10, 18].

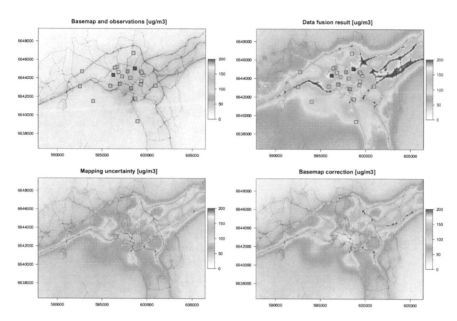

**Fig. 3.10** Data fusion example with real-world observations in Oslo on 6 January 2016 at 09:00 UTC [18]

### 3.3.2.3   Big Data Aspects

In CITI-SENSE, we implemented large-scale engagement from schools (seven secondary schools and 17 elementary schools), universities (3), kindergartens (54), tenant associations (9) and citizens, and performed 9.4 million observations in nine city locations, including: (1) 324 AQMesh air sensors units in network at one time; (2) 327 LEOs unit volunteers; (3) 1200 CityAir App users; (4) 50 volunteers to monitor environmental quality in four public spaces; (5) 2036 reported perceptions by using CityAir App; (6) 1530 answers to air quality questionnaires, and (7) 300 evaluations of COT. The big data aspects of sensor data streaming, and managing the velocity of sensor data is handled by delegating the first storage step to sensor platform dedicated storage services, with a subsequent storage in the common WFS server supporting the CITI-SENSE data model. Figure 3.11 shows the main components involved in running the LEOs sensor unit [6]. A smartphone with ExpoApp2 application installed, was used to read data from the LEOs sensor unit using Bluetooth and upload the data to Ateknea's sensor platform using WI-FI/3G/4G. Both the LEOs sensor unit and the ExpoApp2 application are capable of storing data locally before uploading data to the common sensor platform whenever a preferred communication is available.

**Fig. 3.11** LEO system
architecture

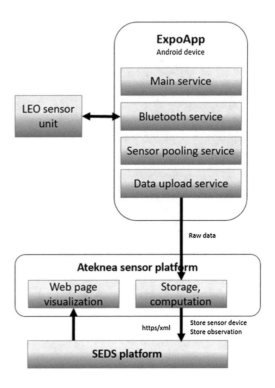

### 3.3.2.4  Reusable Widgets

Reusable widgets have been developed to visualise the collected observations of
sensors and real-time data coming from heterogeneous sources on maps, charts and
plots in web portals and mobile devices, typically based on time-based and location-
based queries. These widgets could be easily configured for and deployed in various
kinds of end user decision making applications and platforms. Currently, all these
reusable widgets are available on CitObs central web portal.[21]

### 3.3.2.5  Web Portals

The Web portals were developed using a content management server called
DotNetNuke (DNN).[22] DNN is an industry leading CMS due to its configurability,
support and $3^{rd}$ part development. DNN comprises a base system, which provides a

---

[21] http://co.citi-sense.eu/CitizensObservatoriesToolbox/Widgets.aspx.

[22] http://www.dnnsoftware.com/community/participate/community-showcase.

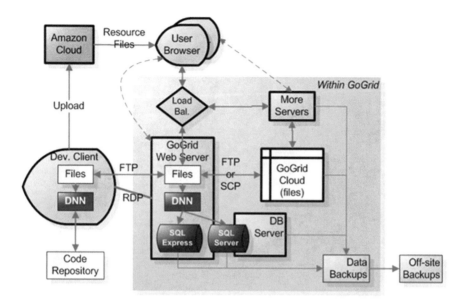

**Fig. 3.12** Citizens' observatories web portals DotNetNuke architecture

website 'out of the box' but can be extended with additional modules, according to need. It is developed using the Microsoft .NET platform. Originally, it was written in VB. NET, though the development has shifted to C# since version 6.0. The Web portals use version 07.04.02 [6]. It runs on MS SQL. It has been configured to provide dynamic content to the users, including text and images, video content, links to social media, as well as access to other similar portals. The DNN architecture is designed to be flexible, scalable and secure. It can utilize the latest storage systems, including cloud storage, as well as traditional storage solutions (Fig. 3.12 [6]).

## 3.4   Examples for Using Citizens' Observatories Toolbox

### 3.4.1   Example 1: Raising Students' Awareness to Air Quality Concepts in Haifa

As part of the project work, Israel Institute of Technology (Technion)[23] in Haifa conducted a campaign in schools to engage students to participate in an air quality monitoring using both portable and static air quality sensors. First, students received

---

[23]http://www.technion.ac.il/en.

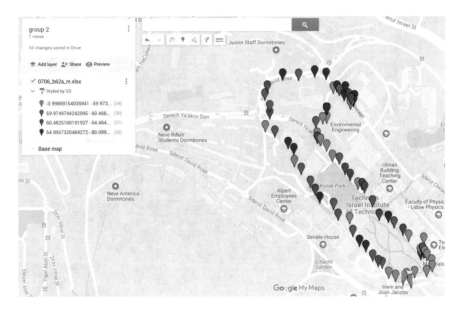

**Fig. 3.13** Track of air quality by using LEOs

detailed maps of the Technion campus, and with guidance of experienced Master and PhD student formed hypotheses of the air quality in different areas of the campus. Students then designed a walking route throughout the Technion campus, and using a portable air quality sensor and a GPS application, monitored the air quality throughout the route (Fig. 3.12). Finally, the walking routes and values of the measured pollutants were presented and discussed, then compared to the initial hypotheses. All routes and values were uploaded for further use of the students and schools. The activity was a great success, as it integrated air pollution knowledge and practice (measurements), many of the participants stated it was the highlight of the project (Fig. 3.13).

In a separate activity, students conducted research projects using air quality data from static air quality sensors installed on the school premises. Two of these works were presented in a regional research-fair at the National Science Museum in Haifa. The two groups examined the difference in air quality in different locations in the school and presented their hypothesis, methods and findings and conclusions. One of the group of students won a certificate of excellence for their work on this relevant and important topic.

To conclude the work done in collaboration with the schools, one of the teachers in-charge wrote a thank you note stating: "In my name and in the name of the whole community, we appreciate and respect the meaningful contribution in raising awareness to air quality issues and in taking responsibility on the environment we live and work in".

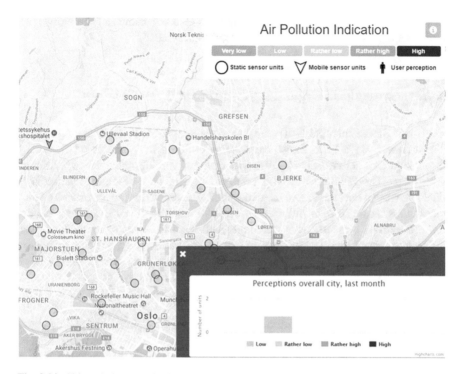

**Fig. 3.14** Citizens' observatories in Oslo

## 3.4.2   Example 2: Citizens' Observatories with Users Being Parents with Children and Regular Citizens in Oslo

In Oslo, static air quality sensors were tested and deployed in 51 kindergartens across 13 areas by the Norwegian Institute for Air Research (NILU).[24] In addition, a number of regular citizens carried portable air quality sensors with them on their daily routines [3]. This data is shown on an interactive map of the city, with the stations coloured according to their current air quality status. Furthermore, there is the option to add additional layers of information, such as the personal perception of air quality, as in this example on the left (Fig. 3.14). The Citizens' observatories with users being parents with Children and regular citizens in Oslo have raised a lot attention from the media and the government. There was an interview for the Norwegian public TV (NRK) in the program "Schrodinger's cat" on 16, January 2015 which reached a high number of the general public in Norway during the prime news time.[25] An article in "AstmaAllergi" about CITI-SENSE and the school

---

[24]http://www.nilu.no.

[25]http://www.nrk.no/viten/slik-unngar-du-skader-fra-lufta-du-puster-1.12151864.

case study was distributed to all the members (15,000 persons) of the Norwegian Asthma and Allergy Association (NAAF) on January 2015.[26]

### 3.4.3 Example 3: Organizing Nature Days for Children

The Jozef Stefan Institute (JSI)[27] in Ljubljana adapted COT tools for organizing "Nature days" in a local forest, introducing children to the topic of air quality through various activities [17]. Different activities were designed for two different age groups. The youngest were 6–7 years old, and the oldest 8–9. In "Nature day", the topic was an introduction to the concept of measurements and uncertainty which researchers tackle in their daily tasks. The children got to learn about air quality through various measurements techniques, where instruments were used, coupled with gaming exercises. The instruments used were a $CO_2$ meter, portable air quality sensor, infrared camera, $1\,m$ stick and a laser distance measurer. In one of these examples, the group was divided in half (around 20 children) where the first half packed in a tent and immediately saw how the $CO_2$ rose, while the other half were outside with an infrared camera (Fig. 3.15).

Introducing gaming was both fun and productive. The children of that age have already developed an idea of the typical pollution sources and how one can influence it. At the end of the "Nature day", the classes were provided with group images taken with an infrared camera, as well as a printed map of air pollution measured in the location covered during that day's activity. Altogether three "Nature days" were organized in the summer 2013 and in 2014. The teachers wished to repeat the concept again the following years. One of the teachers wrote an article about the "Nature day" in Naravoslovna solnica: a journal for teachers of science subjects, sharing practices and inspiring other teachers in Slovenia.

## 3.5 Conclusion

The CITI-SENSE COT facilitated environmental health governance and the results from the seven COT areas of Methods, Data, Web Portals, Widgets, Sensors, Surveys and Mobile Apps have potential to be used in future CitObs projects. **Methods** provided an overview of methodologies and methods developed in the project with references to the relevant documents with a focus on a data model and a data storage service, and data fusion methodology to add value to the observations of pollutants from low-cost micro-sensors. **Data** is one of the main outcomes from the CitObs project, with a particular focus on how to set up a WFS server and the actual data that has been collected through the CitObs project. The CitObs

---

[26]https://www.naaf.no/globalassets/x-gamle-bilder/documents/1.-astmaallergi/aa-1-15/sider-fra-astma1_15_citi-sense.pdf.

[27]https://www.ijs.si/ijsw/JSI.

**Fig. 3.15** The map from an infrared camera

architecture aims at the support of multiple types of sensors and mobile apps for collecting data, and the support of multiple ways of providing data for further use and processing, with the use of a common data model and WFS storage support for the CitObs data. **Web portals** represent the most visible interface to the CitObs results and data, and can be reused to provide access to the similar CitObs data through a map interface. **Widgets** are reusable user interface components that can be reused in future CitObs. The widgets are based on HTML5 to enhance the support for portability across platforms, including maps with sensor locations and index values, real-time and historical sensor values, physical activity level maps and graphs, thermal and acoustic measurement graphs and widgets for questionnaires. **Sensors** and sensor platform development has been a main focus in the CitObs project, with two principal ways for sensor platforms to interact with the WFS server: either through a push (API-based) or a pull (file-based) interface. **Surveys** provide a good approach for interaction with citizens. CivicFlow is a service that can support surveys both from a smart phone and a web portal. The widget framework for Questionnaires by U-Hopper is the basis for interaction with the survey services. **Mobile Apps** are important in order to support citizen's participation. The SENSE-IT-NOW and CityAir Apps are presented as examples of the approach for the development of Mobile Apps. From a developer's point of view the code is available as open source through a GitHub repository. Further work to abstract and generalise the apps as a basis for a more reusable app framework is in progress.

All the tools in the COT have been tested and evaluated in the nine city locations, and were made available online[28] upon completion of the project in 2016. Some of the tools are proprietary and belong to SMEs (Small and Medium-sized Enterprises) and partners within the project. These tools have been extended or updated based on our internal and external requirements, or further developed according to new

---

[28]http://co.citi-sense.eu.

requirements to align with the aim of the project. In some cases, the development code is open source and available to the public. However, this is not applicable to the entire COT. For example, applications like the ExpoApp, CivicFlow, or parts of algorithms and procedures, are not available for reuse. But we do have open source code and products which include (1) data visualization web pages and widget code; (2) SENSE-IT-NOW: Cross platform smartphone application for environmental monitoring toolkit for public places; (3) CityAir: Cross platform smartphone application for collecting citizens air quality perceptions. These applications are freely available for reuse within other projects or initiatives; and (4) Data fusion and data assimilation: a methodology to use crowdsourced observations of air quality for deriving high-resolution urban-scale air quality maps.

The tools presented here evolved from work undertaken in the context of studies funded under the project CITI-SENSE. CITI-SENSE is a collaborative project partly funded by the EU FP7-ENV-2012 under Grant Agreement No. 308524. We want to thank the members of CITI-SENSE, who contributed to the development of the COT.

# References

1. Berre, A.I., Joglekar, B., Fernandez, A., et al. (2016). Deliverable D7.6 - Part 3: Citizens' Observatory Toolbox – Developer perspective. Available at http://co.citi-sense.eu/TheProject/Publications/Deliverables.aspx (accessed on 9th August 2017).
2. Borrego, C., Costa, A.M., Ginja, J., et al. (2016). Assessment of air quality microsensors versus reference methods: The EuNetAir joint exercise. Atmospheric Environment 147: 246–263.
3. Castell, N., Dauge, F.R., Schneider, P. (2017). Can commercial low-cost sensor platforms contribute to air quality monitoring and exposure estimates? Environment International 99: 293–302.
4. Castell, N., Kobernus, M., Liu, H.-Y., Schneider, P., Lahoz, W., Berre, A.J., Noll, J. (2015). Mobile technologies and services for environmental monitoring: The Citi-Sense-MOB approach. Urban Climate 14(3): 370–382.
5. Fishbain, B., Lerner, U., Castell, N., et al. (2017) An evaluation tool kit of air quality micro-sensing units. Science of the Total Environment 575: 639–648.
6. Fredriksen, M., Bartonova, A., Kruzevic, Z., Kobernus, M., Liu, H.-Y., Santiago, L., Schneider, P., & Tamilin, A. (2016). CITI-SENSE. Final report on methodology. Deliverable 6.4, Work Package 6 (NILU report, 29/2016). Kjeller: NILU.
7. Grossberndt, S., and Liu, H.-Y. (2016). Citizen participation approaches in environmental health. In Pacyna, J., & Pacyna, E. (Eds.), Environmental Determinants of Human Health (pp 225–248). London, UK: Springer.
8. Jovašević-Stojanović, M., Bartonova, A., Topalović, D., et al. (2015). On the use of small and cheaper sensors and devices for indicative citizen-based monitoring of respirable particulate matter. Environmental Pollution 206: 696–704.
9. Kotsev, A., Pantisano, F., Schade, S., Jirka, S. (2015) Architecture of a Service-Enabled Sensing Platform for the Environment. Sensors 15: 4470–4495.
10. Kotsev, A., Schade, S., Craglia, M., Gerboles, M., Spinelle, L., Signorini, M. (2016). Next Generation Air Quality Platform: Openness and Interoperability for the Internet of Things. Sensors 16(3): 403.
11. Liu, H.-Y., Bartonova, A., Berre, A., et al. (2015). Deliverable D 4.3 – CITI-SENSE Citizens' Observatories – Version 1. Restricted to the CITI-SENSE consortium.

12. Liu, H.-Y., Kobernus, M. (2016). Deliverable D 4.5 – The CITI-SENSE Citizens' Observatories Web Portal. Available at http://co.citi-sense.eu/TheProject/Publications/Deliverables.aspx (accessed on 8th August 2017).
13. Liu, H-Y., Kobernus, M. (2017). Citizen science and its role in the sustainable development. Chapter in the book with title "Analysing the Role of Citizen Science in Modern Research (Advances in Knowledge Acquisition, Transfer, and Management)". Ceccaroni, L, and Piera, J (Eds). IGI Globe.
14. Liu, H-Y., Kobernus, M., Broday, D., and Bartonova, A. (2014). A conceptual approach to a citizens' observatory - supporting community-based environmental governance. Environ. Health 13:107.
15. Moltchanov, S., Levy, I., Etzion, Y., Lerner, U., Broday, D.M., Fishbain, B. (2014). On the feasibility of measuring air pollution by wireless distributed sensor nodes. Sci Total Environ. 502C: 537–547.
16. Montargil, F., Santos, V. (2017), Citizen Observatories: Concept, Opportunities and Communication with Citizens in the First EU Experiences. In: Paulin A., Anthopoulos L., Reddick C. (eds) Beyond Bureaucracy. Public Administration and Information Technology, vol 25. Springer, Cham.
17. Robinson, J., Kocman, D., Smolnikar, M., Mohorčič, M., Horvat, M. (2014). Empowerment in practice - insights from CITI-SENSE project in Ljubljana. Geophysical Research Abstracts. Vol. 16, EGU General Assembly, EGU2014-5458-1, 2014.
18. Schneider, P., Castell, N., Lahoz, W., Vallejo, I. (2016). Deliverable 6.6 – Data fusion of crowd-sourced air quality observations and dispersion model data for urban-scale air quality mapping. Available at http://co.citi-sense.eu/TheProject/Publications/Deliverables.aspx (accessed on 8th August 2017).
19. Slørdal, L. H., Solberg, S., and Walker, S. E. (2003). The Urban Air Dispersion Model EPISODE applied in AirQUIS 2003, Technical description, Norwegian Institute for Air Research, NILU TR 12/03, Kjeller, Norway, 2003.
20. Wehn, U., Evers, J. (2015). The social innovation potential of ICT-enabled citizen observatories to increase eParticipation in local flood risk management. Technology in Society 42: 187–198.

# Chapter 4
# A Real-Time Streaming and Detection System for Bio-Acoustic Ecological Studies After the Fukushima Accident

**Hill Hiroki Kobayashi, Hiromi Kudo, Hervé Glotin, Vincent Roger, Marion Poupard, Daisuké Shimotoku, Akio Fujiwara, Kazuhiko Nakamura, Kaoru Saito, and Kaoru Sezaki**

**Abstract** Acoustic ecology data have been used for a broad range of soundscape investigations. Counting sounds in a given soundscape is considered an effective method in ecology studies that provides comparative data for evaluating the impact of human community on the environment. In 2016, Kobayashi and Kudo collected a particularly valuable dataset containing recordings from within the exclusion (i.e., difficult-to-return-to) zone located 10 km from the Fukushima Daiichi Nuclear Power Plant in the Omaru District (Namie, Fukushima, Japan). These audio samples were continuously transmitted as a live stream of sound data from an unmanned remote sensing station in the area. In 2016, the first portion of their collected audio samples covering the transmitted sound recordings from the station was made available. Such data cover the bioacoustics in the area. This paper describes the methodologies by which we processed these recordings, in extreme conditions, as preliminary eco-acoustic indexes for demonstrating possible correlations between biodiversity variation and preexisting radioecology observations. The variations in some of these vocalizations were also studied.

H. H. Kobayashi (✉) · H. Kudo · K. Nakamura · K. Sezaki
CSIS, The University of Tokyo, Tokyo, Japan
e-mail: kobayashi@csis.u-tokyo.ac.jp

H. Glotin · V. Roger · M. Poupard
AMU, University of Toulon, UMR CNRS LIS, Marseille, France
e-mail: herve.glotin@univ-tln.fr

D. Shimotoku · K. Saito
GSFS, The University of Tokyo, Tokyo, Japan

A. Fujiwara
GSALS, The University of Tokyo, Tokyo, Japan

© Springer International Publishing AG, part of Springer Nature 2018
A. Joly et al. (eds.), *Multimedia Tools and Applications for Environmental & Biodiversity Informatics*, Multimedia Systems and Applications,
https://doi.org/10.1007/978-3-319-76445-0_4

## 4.1 Introduction

According to a report describing the Chernobyl nuclear disaster penned and published by the International Atomic Energy [1], it is academically and socially important to conduct ecological studies focused on ascertaining the levels and effects radiation exposure has had on wild animal populations over several generations. Although numerous studies and investigations have been conducted regarding the Chernobyl nuclear disaster, there were very few captured audio samples available. In 2012, over 25 years since the Chernobyl disaster occurred, Cusack published audio recordings captured from within the exclusion zone in the Ukraine [2]. To understand the impact a nuclear disaster or other such catastrophic event has on wildlife, we first need long-term and wide-range monitoring of the effects nuclear radiation has on animals because there is little evidence of the direct effects of radioactivity on wildlife in Fukushima [3].

Immediately following the Fukushima Daiichi Nuclear Power Plant disaster, Ishida, a research collaborator at the University of Tokyo, started conducting regular ecological studies of wild animals in the northern Abukuma Mountains where high levels of radiation were detected. In [3], Ishida noted that it is essential to place automatic recording devices (e.g., portable digital recorders) at over 500 locations to properly collect and analyze the vocalizations of the target wild animals. To monitor such species, an effective method is for experts to count the recorded voices of animals; here, acoustic communication is used by various animals, including mammals, birds, amphibians, fish, and insects [4], thus a broad range of species may be covered using this technique. This audial method, in conjunction with visual counts, is commonly used to investigate the habitat of birds and amphibians [4]. It is often surprisingly difficult to analyze this recorded data, which requires observers to manually count and list identified species via repeated playbacks.

Given the intensity of such work, it is also very difficult to continue these activities for long periods of time. Therefore, in this study, we aim to establish a long-term continuously operating ubiquitous system that delivers and analyzes, in real time, environmental information, including bio-acoustic information, for citizen scientists. More specifically, in this paper, we discuss the development and evaluation of an implementation of this system used as part of a bio-acoustic ecology study of the Fukushima accident. Based on related work, we first developed a real-time streaming and identification system for this study, then designed a new experimental human-computation system based on related studies and methodologies. We discuss the methodologies we use to process these recordings as ecoacoustic indexes that demonstrate the variations in biodiversity. Note that while this study is not intended to provide scientific insight into the Fukushima accident, it does provide a comparable dataset and multimedia system for further bio-acoustic studies in the area.

## 4.2  Background

As introduced above, in ecology studies, it is often desirable to develop a multimedia system that most effectively supports a study with minimal resources. Recently, the use of the Internet and cloud-sensing has engaged "citizen scientists" [5], i.e., members of the public that are motivated to work together with professionals interested in science-based conservation to expand the shared knowledge base and explore large-scale solutions. While this has the potential to move science in a good direction, it is difficult to rely largely on citizen participation alone. More specifically, through the cooperation of citizen scientists, activities were also conducted to analyze data actually recorded in restricted area (not exclusion zone), but there is a problem of continuity [6]. Instead, of asking participants to work directly, those that use the knowledge that the participants entered by accident by keeping the system running all the time. We suggest here that this could solve the continuity problem, and we have conducted the research described herein to realize this.

### Developing a System That Can Be Operated for Long Periods of Time

Ecological studies of the environment near urban areas are now being conducted using cell phones [7]; however, it is difficult to use such information devices within an exclusion zone because these areas do not tend to have reliable infrastructure and can be dangerous areas for individuals to enter. Therefore, we conclude that it is necessary to develop a remotely controlled monitoring and evaluation system capable of operating for multiple years to ensure long-term stability under unmanned operating conditions. Note that we previously researched and developed proprietary systems that deliver and record remote environmental sounds in real time for ecology studies [8]. Our previous system was continuously operational on Iriomote Island (Okinawa, Japan) from 1996 to 2010. To date, the fundamental research performed at Iriomote Island has expanded into the Cyberforest project that we are conducting at the University of Tokyo [9].

### Observing User Behavior Via Our Developed System

We have a record of casual comments and analysis from over 2000 users who wrote during experiments we conducted from 1996 to 2010 [8]. Among the conditions that keep environmental sounds running in real time, it was revealed that the user most consciously was the animal's voice. This implies that if we continue to stream real-time environmental sounds to users who are interested in environmental issues, these users will share the names of bark animals with others. Moreover, despite being in situations in which users do not know when an animal will make any noise, users continue to listen carefully until an animal makes sound, then carefully report it. Note that in most of these cases, we are not asking the users to do anything specific; they are taking these actions on their own accord. Given these behaviors, we felt that if we could evaluate these activities performed by citizen scientists, we could solve the aforementioned continuity problem.

**Developing An Interface That Makes it Easy to Obtain Comments from Users**
The development of a wearable forest [10] and a tele echo tube [10] was meant
as a work of art that demonstrates its effect. Here we placed a speaker next to a
microphone already installed in the forest, then observed the reaction by adding
cue sounds to environmental sounds and user actions. From the exhibition of this
artwork, it became clear that the concentration of the user's sound is high. Based on
these findings, we also conducted research on platform for Citizen Science [5]. More
specifically, we are working with ecology scientists to develop a new type of bird
census method, i.e., an audio census, that uses our live streaming audio and social
media systems (e.g., Internet Relay Chat and Twitter) [9]. When ornithologists in
separate locations used our developed sound system to remotely conduct a woodland
birds census with the cue sounds, more species were identified than from a field-
based spot census (i.e., 36 identifications versus only 28, respectively). Given the
above issues, we summarize our goals via the problem statements listed below.

1. Social Problem: It is academically and socially important to conduct ecological
   studies focused on ascertaining the levels and effects radiation exposure has had
   on wild animal populations over several generations in Fukushima. Understanding of research activities from society is important.
2. Technical Problem: Since there are limitations on the working hours and abilities
   of researchers, including both professionals and citizen scientists, it is necessary
   to improve the efficiency of work to the extent possible by utilizing artificial
   intelligence (AI) techniques.
3. Computational Problem: Since it requires a certain level of expertise and time to
   create proper training data, it is necessary for anyone to be able to make training
   data as efficient as possible. It is necessary to clarify the design theory to obtain
   highly accurate data when using unsatisfactory training data.

## 4.3  Developed System

For this study, we installed the first transmitter station [11] within the exclusion
zone area shown in Fig. 4.1; more specifically, this location was 10 km from
the Fukushima Daiichi Nuclear Power Plant. The transmitter station is located

**Fig. 4.1** (**a**) Microphone, (**b**) sync node station and (**c**) website project site in exclusion zone,
which is 10 km from Fukushima Daiichi Nuclear Power Plant

within the Oamaru district in the town of Namie, Fukushima (i.e., 37°28′4.3″ N 140°55′27.5″ E). We selected this site within the exclusion zone because it is one of the most difficult areas for long-term continuous investigations. Here, the exclusion zone is the most radioactively polluted zone in Fukushima. Further, no remote sensing methods are available on the surface due to the lack of power, information, and traffic infrastructures. Given these restrictions, although field surveys are required, the number of workable hours is extremely limited due to radiation exposure concerns. Finally, frequently used portable recorders require regular replacement given their limited memory and battery capacities, which is impractical for long-term continuous investigations.This project development began after a year of literature searching from 2011 to 2014. A transmitter station with satellite Internet was installed in 2015. We then conducted an intensive feasibility survey of all transmission facilities at the location. Finally, after official preparations and approvals, we signed a service contract with a local electricity company in 2016. The final construction was completed at the end of March, 2016 [11].

In this project, we set out to collect, share, and analyze soundscape data within the exclusion zone. At the first, we developed both a Live Sound System and a Streaming/Archiving System that enabled us to distribute sound data from the exclusion zone to the public via the Internet to make such data publicly available for listening in real time via http://radioactivelivesoundscape.net/ and for asynchronous listening. The Live Sound System was composed of separate subsystems, i.e., a Field Encoding System to digitize live sounds from the forest and a Streaming/Archiving System to deliver the live sound data via the Internet and archive the sound data in a recorded file. Note that the technical operational testing notes of the Live Sound System were discussed previously in [11].

The Field Encoding System was composed of two key components, i.e., an audio block and a transmission block. Microphones (omnidirectional SONY F-115B) were individually connected to an amplifier (XENYX 802, Behringer) of the audio block, and their outputs served as input to an audio encoder (instreamer100, Barix) that converted sounds captured by a microphone into MP3, which was the format used for subsequent digital sound delivery. These characteristics of our Internet service plan were important considerations since research funds required to conduct such a long-term ecological study are likely to fluctuate over time. As there was no prior Internet connection at the exclusion zone site, we used a satellite Internet service, which was provided by IPSTAR in April 2016.

The Streaming/Archiving System is located in the server room of our laboratory and uses a normal bandwidth Internet connection, allowing simultaneous public access to transmissions in Fig. 4.2. We employed two servers; one for streaming, the other for archiving. The processed audio signal sent from the microphone was encoded into an MP3 live stream in the Field Encoding System. After transfer to the Streaming/Archiving System, the MP3 live stream can be simultaneously played on MP3-based audio software worldwide. The operating system was the standard single package of Linux Fedora 14, and the sound delivery service was implemented in Icecast 2 software.The servers were established in our laboratory rather than at the contaminated forest site. This setup avoids the technical difficulties in providing power and adequate data download at the remote location.

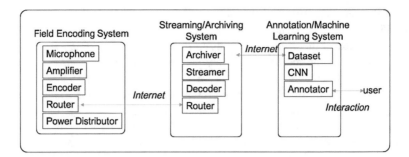

**Fig. 4.2** System diagram of live sound system: field encoding and streaming/archiving system

## 4.4 Data Processing

The recording was conducted from June 2016 to June 2017 in Japan Standard Time in the exclusion zone of Fukushima, Japan. The summer and winter seasons in this zone last from June to September and from December to March, respectively. The average monthly temperature is highest in August (23.0 °C and lowest in January (2.0 °C). The average annual rainfall is 1511 mm (Namie Meteorological Station; JMA 2017) [12].

The audio files were processed by peak normalization high-pass filtering with a 500-Hz cutoff and 20 dB/decade attenuation. These calculations are made with the software *sox* version 14.4.2.

To process the (24 h $\times$ 365 days) sound stream recordings of the environment surrounding the station, the human computation must be augmented with automated analysis. Both are presented below.

**Manual Detection**
In this first strategy of manual annotation sampling, as the acoustic activity of birds is highest near sunrise, we analyzed the live streaming audio data between 10 min before sunrise and 60 min after sunrise [12]. Mainly following the procedure in [13], we also studied the sunset sounds. Audio files containing the sunrise and sunset times were annotated. The recordings were started at 06 min of every hour and lasted for 1 h (i.e., streaming events were recorded at 00:06–01:05, 01:06–02:05, and successive hours throughout the day). Denoting the sunrise or sun-set time by $n{:}m$, if $6 \leq m < 36$, we take the first 30 min of the sound file containing the sunrise or sunset time. Otherwise (i.e., if $0 \leq m < 6$ or $36 \leq m < 60$), we take the last 30 min of the file. To alleviate the workload of the listeners, these 30 min audio files were separated into two parts.

In this experiment, 21 students were recruited to index the selected audio stream. The students were instructed to identify four events: the songs and calls of the target birds, rain, and wind. They were also instructed to subjectively identify the signal levels of each event (strong or weak). The selected target bird was the Japanese Nightingale (*Horornis diphone*) for comparison with Ishida et al. [14].

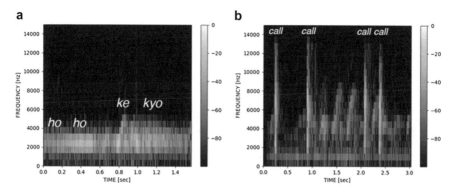

**Fig. 4.3** Vocal activity of bush warblers: colors describes signal strength in dB Full Scale. (**a**) warblers song. (**b**) warblers call

Moreover, the songs of nightingales are unique and common, so their directions are more easily detected than songs of other bird species. The *ho ho ke kyo* songs of male Japanese Nightingales attract females in Fig. 4.3a, and the *chi chi chi* songs of both sexes warn against predators or herald the presence of one bird to others. For this reason, calls are harder to localize and identify than songs in Fig. 4.3b.

To minimize the human error factor, each audio file was scanned at least three times by different listeners. Listeners dedicated 4 h to this task, and were directed to index as many files as possible. Forty-eight audio files were allocated to 21 listeners (11 listeners for the former part, 10 listeners for the latter part). This human listening experiment accumulated 2225 events, including 711 calls, 572 songs, 628 winds and 314 winds. From these human-based inputs, 8006 MP3 files were computed. The CNN detector depicted in next section has been trained on these annotations.

A second strategy of human annotation has been conducted. To build efficiently the Received Operation Characteristics (ROC), specialists made binary annotation for 2500 files, 4 per day at sunrise, sunset, mid day and mid night. For this purpose the DYNI team designed a web-based application for collaborative audio annotation based on a front-end developed by Cartwright et al. [15]. This application makes the annotation of large amount of audio files easier and more robust by allowing the (customizable) integration of annotations from an unlimited number of users. This makes it a useful interface to build large ground truth and train machine learning algorithms, within a citizen science scope. A demo of the system can be found at http://sabiod.org/EADM/crowdannot and is proposed with open licence for academic research. The interface is presented in Fig. 4.4. We then produced in 5 h an expert annotation of 2500 chunks of 10 s. In Toulon 4 professionals labeled for existence of sound sourced by a bird, and in Tokyo 2 professionals labeled for sound by a warbler. Regarding these data as a ground truth, the detector model performance was evaluated.

**Automatic Detection**

For the automated analysis we tried two methods: (1) a raw signal processing and (2) a deep neural net approach. The signal processing approach is based on energy

**Fig. 4.4** Illustration of the online collaborative annotation system we developed to annotate 40 s of recordings per day with annotator based in Japan and in France, in order to compute the ROC of two detectors and other statistics

and spectral flatness [16] thresholding. The threshold for local decision were guided by the annotated data and Receiving Operating Characteristics analysis.

With the second automatic method, the neural net, the audio files were splitted and differentiated by windows of 0.25–2 s with 50% overlap. To increase the robustness of the model, the converted raw files were augmented by added noise, then fast-Fourier transformed was computed as the input of a 8-layers Convolutional Neural Network (CNN). A denoising auto-encoder was firstly trained to initialize the convolutional layers. The number of connections in this architecture was 32–512 [17]. The parameters of each layer (weights and biases) were L1/L2 regularized (multiple combinations and hyper-parameters were tried). We tried multiple combinations of conduction for the convolution layers (maximum pooling, average pooling and batch normalization).

## 4.5 Results

The recordings worked fine, yielding to a full year, 7/24 soundscape recording, and nearly 2 To of data. We trained the CNN on the first annotation strategy, and ran it at a scale of 0.5 s. on the whole year. The Accuracy of the CNN model was poor (under 30%). The reason could the hudge amount of rain noise, and the high variability of the used annotations for the training stage.

We then also score the automatic detector based on the maximum of the spectral flatness on the 10 s. sections. The ROCs of this detector are given Fig. 4.5 for Bird and for Warbler detectors. Even if it is a very fast annotator running in few hours the

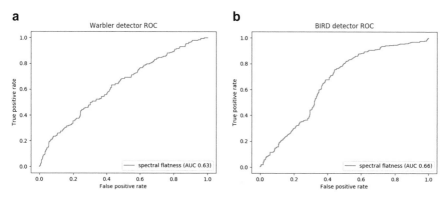

**Fig. 4.5** ROC curve of (**a**) Bird detector, and (**b**) Warbler detector based on spectral flatness

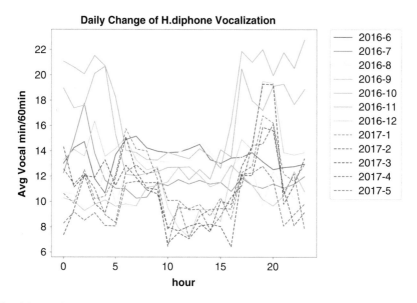

**Fig. 4.6** Monthly vocalization change of warblers: hourly-average is illustrated corresponding to each o'clock. Legends describes years-months

whole year, the area under the curve for Bird detector is 0.63 and 0.66 for Warbler, which is allowing preliminary analyses depcited below.

At 0.5 s intervals, the approach outputted whether the sound included a call or a song (a warble), but did not distinguish between calls and songs. The model counted 6520 h of warbling. The vocal activity of warblers was most variable at dawn and dusk in Fig. 4.6.

Corresponding to temperature and season, two groups of warbler detection can be classified, demonstrating that the model allow to produce ecological features that could be correlated at long term (Fig. 4.7).

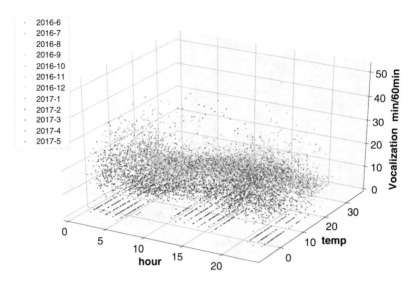

**Fig. 4.7** Numbers of warbling vs. hours and temperature: warbling count is shown with o'clock-time of a day and temperature. Legends describe years-months

## 4.6 Discussion

The raining conditions (weak signal) and difficulties on the field to increase the recording capacity yield to difficulties into the automatic detections with trained CNN. However, we demonstrated that we get reasonable eco-acoustic results all over the year.

Vocal activity was identified during periods of heavy rain ($\geq 20$ mm/h), when warbling is known to cease. As evidenced by the lack of target sounds in their assigned audio files, human's attention lapses, the signal can be unrecognized by the human. When simultaneous signals arrive from multiple directions, they will overlap, which complicates the computation. In contrast, humans can focus on a specific direction (the so-called cocktail party effect), if multiple microphones are installed at the same site. If a bird is present but does not sing, it will be unrecognized by both humans and the model. This problem could be overcome by a visual observation method; for instance, the Cyberforest Project [9] is installing a monitoring camera at their observation site.

The difficulty in the automatic method yields in the weak signal of the target (under rain). Recently the Bird activity detection challenge confirmed the superiority of CNN approaches as the model of Pelligrin [17]. Also CNN is the best model for Bird species classification [18]. These CNN-based systems required supervisors annotated by professional ornithologists, however creating supervisors is very time consuming task that it is not possible to rely completely on experts [6]. Asking amateurs with soundscape mania can help this problem. This study demonstrated limitations and possibilities of connecting human-based annotation to machine learning systems. Future work will focus to obtain highly accurate annotated samples using the produced interface, or potentially recruiting massive international crowd sourcing as the interface allows cooperative annotation. Detection of humans at the training stage are not always suitable as computer inputs. Multiple calls are always heard at any one time. Whereas some listeners record these calls as separate events, others count them as a group in Fig. 4.3b. Other than that factor, the limited proficiency of listeners in using the listening software, which degraded listening precision 1/1000–1/10 s. To control and overcome such problems and to accelerate the experiment, united interface between users and CNN is desired. To overcome these limitations, we augmented the raw data and applied filtering thresholds on the features to maximize the area under the curve with labeled data. Machine learning can over-estimate the number of vocal activities of bush warblers, whereas human listeners cannot always distinguish between bird warbling and the calls of frogs, and may combine both sounds into the learner.

As previously mentioned, it is necessary to improve the efficiency of work to the extent possible by utilizing semi-supervised automatic algorithm for the professionals and citizen scientists.

We however demonstrated the presence of nightingales near the station varied on a monthly basis between June 2016 and July 2017. The acoustic index was elevated during the breeding season in spring, when male birds sing frequently to attract females. This study commenced in 2016, and because no comparative record exists before the Nuclear Accident, the effect of radioactive contamination cannot be assessed. The dataset collected at Namie, Fukushima, is the largest dataset available for ecoacoustic research. Using this dataset, we developed a methodology for tracking environmental effects on biodiversity. The dataset is also useful for measuring and comparing future long-term changes. As the recording has continued every 24 h since June 2017, we can analyze the yearly changes from this time onward. The Cyberforest project operates eight other observational open-microphone sites located throughout Japan. All the sites are connected to the on-line archive system at http://www.cyberforest.jp/. Applying this method at the other sites, we could simultaneously compare the bio-acoustic data at multiple sites. This technology could monitor climate change through birdsong analyses.

There is little known in advance about the introduction and long-term operation of consumer electronics products in high-dose zones. Long-term operation for decades in this project has many technical problems. Especially, although the report is informal, it has been reported that the operation time with batteries is shortened when consumer electrical appliances are operated in a high dose zone. In other words, it means that the brought-in equipment can not operate for the prescribed time at first. However, since it is not permitted to bring radioactive equipment out of the same zone by operating in a high dose zone, research is necessary for actual verification. Since the equipment used in this study is already highly radioactive, it becomes a sample for such experiments.

As previously mentioned, it is academically and socially important to conduct ecological studies focused on effects radiation exposure has had on wild animal populations over several generations in Fukushima. (Social Problem)

Although 6.5 years has passed since the Great East Japan Earthquake, the situation in the difficult-to-return area has remained static. Electricity remains in disrepair, and many patches of road are still collapsed. As decontamination activities are rarely performed, it is extremely difficult for researchers to enter the site. Building a multimedia system at such a point and using it for a long time is not only technically but also a socially big challenge. Since the power supply and road collapsed due to the earthquake, it can not be expected to be a base station of mobile phones or a stable power supply. Special permission is required for the surveyor to conduct the ground survey directly and the staying time is extremely limited due to the high dose. However, due to the large environmental problem of nuclear disaster, there are high social concerns about the animals left here. In order to make full use of this limited resource, this research is aimed at constructing a multimedia system capable of long-term operation, taking advantage of environmental concerns in the general society, It is to realize support. The system constructed for this research has operated continuously for 1.5 years. The system is designed to operate for 24 h over 365 days each year. Prior to this project, we operated a similar system for more than 10 years [8]. Therefore, the present project is expected to continue until approximately 2030. By special arrangement with the power company, it also receives a stable power supply which is unlikely to disconnect in future operation. After finishing the operation of the mp3 file, we will commence operations on the uncompressed recordings.

The National Institute of Radiological Sciences have confirmed plant malformations and other abnormalities at the site [19], we are also considering a camera-based study. Plant malformations are known to be short-term events, but could be recorded by a camera system running for 24 h. Moreover, if individual animals can be discriminated in the camera data and the relevance of deformed plants assessed, we could begin to research the linkage between animals and plants.

Citizens can contribute to ecological surveys by participating in identification activities and research at dedicated events. Experience programs and exhibitions are open to the public at museums, science museums, and similar institutes. Asking participants directly for such work would certainly assist science, but listening to mass data is mentally demanding and probably unsustainable. The present research takes an indirect approach. Our method processes a large amount of data by AI without compromising the creation of training data. The following questions remain to be addressed:

1. How can we request a user to create training data that are easily handled by AI? The training data must be appropriately selected for the total data size. For instance, birdsongs are often sampled in the morning and evening, when vocal activity is most intense.
2. How can we handle differences among individuals in human-generated training data for AI processing? When preparing training data, it is necessary to resolve the timing deviations among users. The timing of a bird chirp varies from user to user. Whereas one person records the moment the bird starts singing, others might record a moment during the singing or when the song has finished.

3. How can we sustain 1 and 2 in a sustainability-oriented society that is increasingly aware of environmental issues? Instead of unilaterally requesting users to create training data, we must develop an interface that promotes environmental consciousness through enjoyable activities and games.

## 4.7  Conclusion

This paper discussed the acquisition and analysis of environmental sounds in a difficult-to-return area 10 km from the Fukushima Daiichi Nuclear Power Station. After obtaining official permission, we established a real-time acquisition and distribution system and acquired over 8000 h of continuous data. To process these data, we used a signal processing approach and tried a CNN model with human-computation. Prior to our study, no samples in this area had been continuously collected over the long term. We believe that by future data acquisition and analysis, we can investigate the influence of radiation on wild animals.

**Acknowledgements** Special thanks go to Fumikazu Watanabe and Laboratory Members. This study was supported by JSPS KAKENHI Grants 26700015 and 16K12666, MIC SCOPE Grants 142103015 and 162103107, JST PRESTO 11012, the Telecommunications Advancement Foundation, the Moritani Scholarship Foundation, the Tateisi Science and Technology Foundation, and the Mitsubishi Foundation. We thank the http://sabiod.org/EADM GDR CNRS MADICS that support development of the annotator platform. We thank Pascale Giraudet for annotation participation and other volunteers in Toulon and in Japan.

## References

1. Chernobyl Forum. Expert Group "Environment.," International Atomic Energy Agency. (2001) Environmental consequences of the Chernobyl accident and their remediation: twenty years of experience; report of the Chernobyl Forum Expert Group 'Environment'. Radiological assessment reports series. International Atomic Energy Agency, Vienna
2. Cusack P (2012) Sounds from Dangerous Places. ReR Megacorp, Thornton Heath, Surrey,
3. Ishida K (2013) Contamination of Wild Animals: Effects on Wildlife in High Radioactivity Areas of the Agricultural and Forest Landscape. In: Nakanishi TM, Tanoi K (ed) Agricultural Implications of the Fukushima Nuclear Accident. Springer, Japan
4. Begone M, Harper JL, Townsend, CR (1990) Ecology: individuals, populations, and communities. 2nd. edn. Blackwell Scientific, Boston
5. Louv R, Fitzpatrick JW (2012) Citizen Science Public Participation in Environmental Research. 1st. edn. Cornell University Press, Ithaca
6. Fukasawa K, Mishima Y, Yoshioka A, Kumada N, Totsu K, Osawa T (2016) Mammal assemblages recorded by camera traps inside and outside the evacuation zone of the Fukushima Daiichi Nuclear Power Plant accident. Ecological Research 31 (4):493–493. doi:10.1007/s11284-016-1366-7
7. Slabbekoorn H, Peet M (2003) Ecology: Birds sing at a higher pitch in urban noise. Nature 424 (6946):267–267

8. Kobayashi H (2010) Basic Research in Human-Computer-Biosphere Interaction. The University of Tokyo, Tokyo, Japan
9. Saito K, Nakamura K, Ueta M, Kurosawa R, Fujiwara A, Kobayashi HH, Nakayama M, Toko A, Nagahama K (2015) Utilizing the Cyberforest live sound system with social media to remotely conduct woodland bird censuses in Central Japan. Ambio 44 (Suppl 4):572–583. doi:10.1007/s13280-015-0708-y
10. Kobayashi H, Hirose M, Fujiwara A, Nakamura K, Sezaki K, Saito K (2013) Tele echo tube: beyond cultural and imaginable boundaries. In: 21st ACM international conference on Multimedia, Barcelona, Spain, pp 173–182. doi:10.1145/2502081.2502125
11. Kobayashi HH, Kudo H (2017) Acoustic Ecology Data Transmitter in Exclusion Zone, 10 km from Fukushima Daiichi Nuclear Power Plant. Leonardo 50 (2):188–189. doi:10.1162/LEON_a_01416
12. Japan Meteorological Agency – Namie Meteorological Station. Japan Meteorological Agency. http://www.jma.go.jp. 2013
13. Ueta M, Hirano T, Kurosawa R (2012) Optimal time of the day to record bird songs for detecting changes of their breeding periods. Bird Research 8:T1-T6. doi:10.11211/birdresearch.8.T1
14. Ishida K, Tanoi K, Nakanishi TM (2015) Monitoring free-living Japanese Bush Warblers (Cettia diphone) in a most highly radiocontaminated area of Fukushima Prefecture, Japan. Journal of Radiation Research 56 (Suppl 1):i24-i28. doi:10.1093/jrr/rrv087
15. Cartwright, M., Seals, A., Salamon, J., Williams, A., Mikloska, S., MacConnell, D., Law, E., Bello, J., and Nov, O (2017) Seeing sound: Investigating the effects of visualizations and complexity on crowdsourced audio annotations. In Proceedings of the ACM on Human-Computer Interaction, 1(1).
16. Ricard, J. and Glotin H (2016) Bird song identification and monitoring system. LSIS internal research report, University of Toulon (ed)
17. Stowell D, Wood M, Stylianou Y, Glotin H Bird detection in audio: A survey and a challenge. In: 2016 IEEE 26th International Workshop on Machine Learning for Signal Processing (MLSP), 13–16 Sept. 2016 2016. pp 1–6. doi:10.1109/MLSP.2016.7738875
18. Sevilla, A, and Glotin H (2017) Audio bird classification with inception-v4 extended with time and time-frequency attention mechanisms. In: 2017 CLEF Working Notes 1866, Dublin
19. Watanabe Y, Ichikawa Se, Kubota M, Hoshino J, Kubota Y, Maruyama K, Fuma S, Kawaguchi I, Yoschenko VI, Yoshida S (2015) Morphological defects in native Japanese fir trees around the Fukushima Daiichi Nuclear Power Plant. Scientific Reports 5:13232. doi:10.1038/srep13232

# Chapter 5
# Towards Improved Air Quality Monitoring Using Publicly Available Sky Images

Eleftherios Spyromitros-Xioufis, Anastasia Moumtzidou,
Symeon Papadopoulos, Stefanos Vrochidis, Yiannis Kompatsiaris, Aristeidis
K. Georgoulias, Georgia Alexandri, and Konstantinos Kourtidis

**Abstract** Air pollution causes nearly half a million premature deaths each year in
Europe. Despite air quality directives that demand compliance with air pollution
value limits, many urban populations continue being exposed to air pollution levels
that exceed by far the guidelines. Unfortunately, official air quality sensors are
sparse, limiting the accuracy of the provided air quality information. In this chapter,
we explore the possibility of extending the number of air quality measurements
that are fed into existing air quality monitoring systems by exploiting techniques
that estimate air quality based on sky-depicting images. We first describe a comprehensive data collection mechanism and the results of an empirical study on the
availability of sky images in social image sharing platforms and on webcam sites.
In addition, we present a methodology for automatically detecting and extracting
the sky part of the images leveraging deep learning models for concept detection
and localization. Finally, we present an air quality estimation model that operates
on statistics computed from the pixel color values of the detected sky regions.

## 5.1 Introduction

Environmental data are crucial both for human life and the environment. Especially,
the environmental conditions related to air quality are strongly related to health
issues (e.g. asthma) and to everyday life activities (e.g. walking, cycling). Thus, it is

E. Spyromitros-Xioufis (✉) · A. Moumtzidou · S. Papadopoulos · S. Vrochidis · Y. Kompatsiaris
Centre for Research & Technology Hellas – Information Technologies Institute, Thessaloniki,
Greece
e-mail: espyromi@iti.gr; moumtzid@iti.gr; papadop@iti.gr; stefanos@iti.gr; ikom@iti.gr

A. K. Georgoulias · G. Alexandri · K. Kourtidis
Democritus University of Thrace, Xanthi, Greece
e-mail: argeor@env.duth.gr; alexang@auth.gr; kourtidi@env.duth.gr

© Springer International Publishing AG, part of Springer Nature 2018     67
A. Joly et al. (eds.), *Multimedia Tools and Applications for Environmental
& Biodiversity Informatics*, Multimedia Systems and Applications,
https://doi.org/10.1007/978-3-319-76445-0_5

necessary to provide citizens with up-to-date notifications regarding environmental conditions. Typically, air quality data are measured by official measurement stations established by environmental organizations and are made available to the public through web sites or web services. However, official stations are few and mainly located in urban areas, thus motivating use of crowdsourcing solutions to improve the geographical coverage and density of air quality measurements. To this end, a number of air quality monitoring initiatives (e.g. luftdaten.info[1]) have emerged that promote the establishment of personal environmental stations by citizens, based on low-cost and relatively easy-to-use hardware sensors. At the same time, the increasing popularity of social media has resulted in massive volumes of publicly available, user-generated multimodal content that can often be valuable as a sensor of real-world events [1]. This fact coupled with the rise of citizens' interest in environmental issues and the need for direct access to environmental information everywhere (both urban and rural areas) and without any extra specialized equipment, has triggered the development of applications that make use of social data for collecting environmental information and creating awareness about environmental issues. In this context, this paper presents a framework that involves the collection of publicly available images from social media platforms and public webcams, their processing using image analysis techniques, and the application of a method for mapping image color statistics to an air quality index. The proposed framework is part of a platform developed by the hackAIR project[2] that gathers and fuses environmental data and specifically particulate matter (PM) measurements from official open sources and from user generated content.

## 5.2 Related Work

Several initiatives attempt to provide citizens with environment-oriented information collected from different data sources. Examples of such initiatives are: (a) iSCAPE[3] that encapsulates the concept of smart cities by promoting the use of low cost sensors and the use of alternative solution processes to environmental problems, (b) the Amsterdam Smart Citizens Lab[4] that uses smartphones, smart watches, and wristbands, as well as open data and DIY sensors for collecting environmental data, (c) CITI-SENSE[5], which provides air quality information based on portable

---

[1]http://luftdaten.info.

[2]http://www.hackair.eu.

[3]http://horizon2020projects.com/sc-climate-action/h2020-making-cities-sustainable.

[4]https://waag.org/en/project/amsterdam-smart-citizens-lab.

[5]http://www.citi-sense.eu.

and stable sensors, (d) CAPTOR[6], which aims at engaging a network of local communities for monitoring tropospheric ozone pollution using low-cost sensors, and (e) AirCasting[7], which is an open-source platform that consists of wearable sensors that detect changes in your environment and physiology, including a palm-sized air quality monitor, an Android app, and wearable LED accessories.

The aforementioned projects use sensors, open data and smart watches as sources. Another source for estimating air quality that has received recently a lot of attention is photos due to their abundance and the fact that no specialized equipment is required. Initiatives that use photos as source for estimating air quality are: (1) the AirTick[8] application which estimates air quality in Singapore by analyzing large numbers of photos posted in the area, (2) the Visibility[9] mobile application that encourages users to upload images of sky to get response regarding visibility which is an indicator of the air pollution of the area and (3) the hackAIR project's air quality platform that combines data from various sources including images posted in social media and retrieved from public webcams.

The AirTick application [28] is a mobile app that can turn any camera enabled mobile device into an air quality sensor. AirTick leverages image analytics and deep learning techniques to produce accurate estimates of air quality following the Pollutant Standards Index (PSI). AirTick first extracts the haziness from a single photo and then converts it into an appropriate PSI value. With haziness extracted from a given image, AirTick passes the haziness information to a Deep Neural network Air quality estimator (DNA) to learn to associate given haziness matrices with PSI values. DNA is designed based on the Boltzmann Machine (BM), which is a neural network of symmetrically coupled stochastic binary nodes. The conducted experiments showed that AirTick achieves, on average, 87% accuracy in day time operation and 75% accuracy in night time operation. Although results are encouraging, a limitation of the AirTick approach is that low light conditions prevent the successful extraction of the haziness component of the images and lead to accuracy deterioration.

Regarding the Visibility application, it is based on the work of [30] that requires users to take pictures of the sky while the sun is shining, which can be compared to established models of sky luminance to estimate visibility. Visibility is directly related to the concentration of harmful "haze aerosols", tiny particles from dust, engine exhaust, mining or other sources in the air. Such aerosols turn the blue of a sunlit clear sky gray. The Visibility app uses the accelerometers and the compass incorporated on smartphones to capture its position in three dimensions while the

---

[6]http://captor-project.eu.

[7]http://aircasting.org.

[8]https://www.youtube.com/watch?v=l11abvYgvBY.

[9]http://robotics.usc.edu/~mobilesensing/Projects/AirVisibilityMonitoring.

GPS data and time are used to compute the exact position of the sun. The system has been tested in several locations in the US, including Los Angeles and Phoenix. However, a drawback of the method is that it requires the images to depict only or mostly sky, thus depending a lot on human judgement. Also users are requested to specify explicitly the part of the image that contains sky pixels which adds considerable manual effort.

Apart from the applications mentioned, several studies were carried out regarding the estimation of air quality from images. In [21], the authors utilize six image features together with additional information such as the position of the sun, date, time, geographic information and weather conditions, etc., to estimate the amount of $PM_{2.5}$ (particles with aerodynamic diameter less than 2.5 micrometers) in the air. Experimental results have shown that the image analysis method is able to estimate the $PM_{2.5}$ index accurately. Nevertheless, the method relies on the manual selection and labelling of the regions of interest in order to operate effectively. This step requires the users to precisely label the buildings in the photos they have taken, which incurs significant overhead. Furthermore, the additional information required by the method on top of the photos and labels of buildings may not always be available, especially in outdoor locations without Internet access.

Another work is that of [46] that proposes an effective CNN-based model tailored for air pollution estimation from raw images. Specifically, the proposed model involves the use of a negative log-log ordinal classifier to fit the ordinal output well, and the use of a new activation function for photo air pollution level estimation. The proposed approach was validated with qualitative and quantitative evaluations on a set of images taken in Beijing against several state-of-the-art methods and it was found to incur smaller error in the air quality estimation.

Finally, in [20], the authors propose a system to estimate haze level based on a single photo. The method proposed involves estimating a transmission matrix generated from a haze removal algorithm, and estimates the depth map for all pixels in the photo. A haze level score is computed by combining the transmission matrix and depth map, and can be calibrated to estimate the $PM_{2.5}$ level. The method was evaluated both on synthetic and real photos providing promising results especially in the synthetic database. Regarding the real photos, further research is required in order to make large scale monitoring based on online user photos more reliable.

Saito and Iwabuchi [32] recently introduced a new method for measuring aerosol optical properties from digital twilight photos. Their method allows for the estimation of tropospheric and stratospheric aerosols, being very promising, despite the fact that it focuses on twilight photos only. Zerefos et al. [44] had previously introduced a simpler approach to retrieve aerosol loadings from paintings from the period 1500–1900. It was found that aerosol concentrations increased in the atmosphere following major volcanic eruptions. These eruptions inserted huge amounts of aerosols in the stratosphere which remained there for years leading to more reddish sunsets. Zerefos et al. [45] extended the research from Zerefos et al. [44], covering the period 1500–2000.

A method close to that of Zerefos et al. [44] is followed in this work to estimate the aerosol load in the atmosphere as described in detail in Sect. 5.6. However, the method is not limited to sunset conditions, is extended to images from users, social media and public webcams and furthermore uses a better representation of the local atmospheric characteristics. The methodology described in this chapter is part of the framework developed within the hackAIR project and constitutes an update of the system presented in [25] that overcomes several of its limitations (e.g. need for more images, better sky localization methods).

## 5.3   Overall Air Quality Monitoring Framework

Figure 5.1 depicts the proposed framework. The framework produces PM measurement estimations using recent (i.e. within the last 24 h) publicly available images. These images are retrieved from media sharing platforms such as Flickr and public webcams. The use of different sources aims to address the need for measurements that are both large in number and cover a large area. Specifically, images retrieved

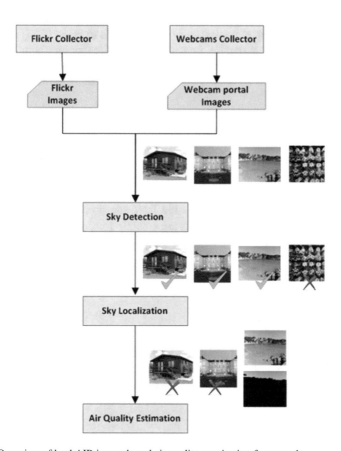

**Fig. 5.1**  Overview of hackAIR image-based air quality monitoring framework

from media sharing platforms offer the advantage of abundance and high geographic coverage (user generated images are expected to be captured in both rural and urban areas) while images coming from webcams offer the advantage of standard delivery of data on a daily basis, with known and standard quality and with fixed location (webcams are usually installed in urban areas). With regard to social media platforms, we use solely Flickr due to API usage restrictions imposed by other popular image sharing platforms that we considered (see Sect. 5.4.1). The other source of images is public webcams that depict parts of the skyline of an area of interest.

After having collected the images from the aforementioned sources, a series of steps is realized that aim at producing air quality estimations. Initially, a sky concept detector is applied that detects whether sky is depicted in the image by using low-level visual features and a classifier. In the sequel, sky localization detects the sky regions within the image. Two approaches are considered for sky localization, one based on deep learning techniques and the other on heuristic rules provided by air quality estimation experts. The methods are used in a complementary way in order to achieve better results compared to the results produced by either of the two approaches alone. The parts of the images that are identified as sky are used for measuring pixel color statistics, specifically the red to green (R/G) and green to blue (G/B) ratios. The last step involves using these ratios for providing information about the aerosol content of the atmosphere, which can be translated to air quality estimation in the form of air quality index (e.g. low, medium, high).

## 5.4 Public Image Collection

Social media platforms and webcam sites constitute the sources for collecting regularly updated publicly available images for the proposed framework. These images should be geotagged to be usable from the proposed framework. In this section we present which social media are suitable for collecting images, how we retrieve data from them, as well as the webcams repositories that include webcams dispersed around the world.

### 5.4.1 Social Media Platforms

Users upload billions of images on a daily basis in social media. However, not all social media are suitable or equally popular for posting images. The KPCB Internet Trends Report 2016[10] provides an overview of the trends related to image

---

[10]http://www.kpcb.com/blog/2016-internet-trends-report.

sharing/posting for 2005–2015. Users upload more than three billion images per day in social networks, and the top platforms for photo sharing are Snapchat, Facebook Messenger, Instagram, WhatsApp and Facebook. Unfortunately, a careful examination of these platforms reveals that Snapchat, Facebook Messenger, and WhatsApp do not distribute the user-contributed images through a free API. Instagram, on the other hand, added in June 2016 strict limitations on the apps that could access the data and the number of data they could retrieve, and finally, Facebook allows access only to images from public pages and not from personal user profiles which significantly limits the number of available images.

According to KPCB Internet Report 2014[11], Flickr is the next social network in terms of image uploads with more than 3.5 million new images uploaded daily in 2013[12]. Flickr provides an open API that enables gathering all public images users share through their profiles. Given the specifications and strict limitations of the other social media platforms as well as the considerable amount of data uploaded to Flickr, we conclude that Flickr is the most appropriate source of publicly available social images.

The Flickr collector periodically calls the Flickr API in order to retrieve the URLs and necessary metadata (e.g. timestamp, geolocation) of images captured within the last 24 h. The collection of geotagged images is conducted by submitting geographical queries to the `flickr.photos.search` API method, using the `woe_id` parameter as input. This parameter allows geographical queries based on WOEID[13] (Where on Earth Identifier), a 32-bit identifier that uniquely identifies spatial entities and is assigned by Flickr to all geotagged images. Moreover, to retrieve only photos taken within the last 24 h, the `min/max_date_taken` parameters of the `flickr.photos.search` endpoint are used, which operate on the image's Exif metadata. For the geographical area of Europe, the Flickr API returns about 5000 geotagged images per day on average.

### 5.4.2  Webcam Image Collector

Another source of sky images is public outdoor webcams. Compared to images from social networks, webcams offer the advantage of providing a continuous stream of images from fixed and a priori known locations. As sources of public outdoor webcams, two large-scale repositories are used, AMOS[14] [17] and webcams.travel[15].

---

[11] http://www.kpcb.com/blog/2014-internet-trends.

[12] https://en.wikipedia.org/wiki/Flickr.

[13] https://en.wikipedia.org/wiki/WOEID.

[14] http://amos.cse.wustl.edu.

[15] https://www.webcams.travel.

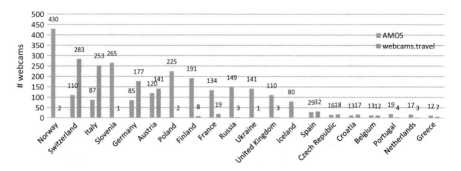

**Fig. 5.2** Geographical distribution of webcams from AMOS and webcams.travel

Based on a set of experiments that we conducted, we found that many of the webcams discovered using standard search engines (e.g. Google or Bing) for a specific location are already contained in either AMOS or webcams.travel. Therefore, we believe that these two repositories cover adequately the needs of the proposed framework and there is no need for a specialized webcam discovery framework. Combined, these sources provide data from more than 25,000 webcams in Europe, which is our main area of interest. Figure 5.2 depicts the geographical distribution of webcams stored in the two repositories (top 20 countries are shown).

### 5.4.2.1 Collecting Images from AMOS Repository

In the case of AMOS, a web data extraction framework was developed that involves downloading and parsing the web page of each webcam and retrieving the images captured within the last 24 h. In order to identify the web page URLs of the webcams located in Europe, we use a search form provided by the AMOS web site that allows performing geographical queries by specifying bounding box coordinates. The number of webcams located in Europe is 4893; however we should note that not all matching webcams are active. The results page is parsed to extract the URLs of the webcams and each page is downloaded and parsed to extract the necessary information. The AMOS image collector is executed four times per day. An analysis of the images collected for a period of 2 months showed that 2246 of the 4893 webcams are active.

### 5.4.2.2 Collecting Images from webcams.travel Repository

Webcams.travel is a very large outdoor webcams directory that currently contains 64,475 landscape webcams worldwide. Webcams.travel provides access to webcam data through a comprehensive and well-documented free API[16]. The provided API

---

[16]https://developers.webcams.travel.

is RESTful (i.e. the request format is REST and the responses are formatted in JSON) and is available via Mashape[17]. In order to collect data from European webcams, an image collector application is implemented that uses the webcams.travel API. In this type of queries the `/webcams/list/` endpoint is exploited along with the `continent=EU` explicit modifier which narrows down the complete list of webcams to contain only webcams located in Europe. Moreover, two implicit modifiers are used: (a) `orderby` which enforces explicit ordering of the returned webcams, and (b) `limit` which is used for slicing the list of webcams by limit and offset. The use of the `limit` modifier is necessary because the maximum number of results that can be returned with a single query is 50. The last part of the query (`show=webcams:basic,image,location`) is used so that in addition to the basic information for each webcam (id, status, title), the returned webcam objects also contain the URL of the latest image captured from the webcam (and its timestamp) as well as the webcam's exact geographical location. Similarly to the AMOS image collector, the webcams.travel image collector is executed four times per day.

### 5.4.3 Image Collection Statistics

The three image collectors, i.e. the Flickr collector, the AMOS webcams collector and the webcams.travel collector, have been collecting images since 24/2/2017, 6/3/2017 and 2/5/2017, respectively. During this period and until 15/5/2017 1,019,938 images had been collected in total across the whole Europe from all sources. Figure 5.3 shows the number of images collected daily from each source. A close examination of the graph shows that the number of images collected each day by the two webcam image sources is almost stable since an almost fixed number of webcams are visited a fixed number of times each day. In particular, 2246 webcams from AMOS and 1000 webcams from webcam.travel are visited exactly four times per day and, as a result, about 9000 and 4000 images, respectively, are collected daily from these sources. On the other hand, the number of images collected daily from Flickr exhibits a large variability since it depends on the number of geotagged images (in Europe) that are uploaded daily by Flickr users. As expected, the number of images collected from Flickr increases significantly during Saturday and Sunday, since users tend to capture and upload more images during weekends. On average, about 5500 images are collected daily from Flickr.

---

[17]https://market.mashape.com/webcams-travel/webcams-travel.

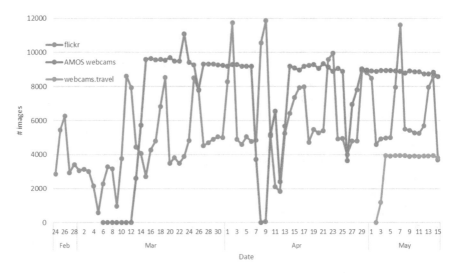

**Fig. 5.3** Number of images collected daily from each source

## 5.5    Image Analysis for Sky Detection and Localization

The next step after image collection is image analysis. This comprises two procedures that are based on sophisticated machine learning and computer vision algorithms; sky detection and sky localization. Given an input image, sky detection is first applied to determine whether sky is depicted in the image, and in case it does, sky localization is applied to determine its exact position (i.e. image pixels). In the sequel, we present an overview of state of the art methods for *concept detection and localization*[18], and then present the proposed framework.

### *5.5.1    Sky Detection*

#### 5.5.1.1    State of the Art

Concept detection in images aims at annotating them with one or more semantic concepts (e.g. sky, trees, road, shadows, etc.) that are chosen from a pre-defined concept list [38]. In general concept detection systems follow a process that first

---

[18]Although in our work we are interested only in the "sky concept", the discussed methods have been designed to work for a wide range of visual concepts and are therefore widely known as concept detection/localization methods.

performs extraction of visual features, then training of classifiers for each concept using a ground-truth annotated training set, and finally, application of the trained classifiers to the features extracted from the unlabeled images that return a set of confidence scores for the presence of the different concepts.

Feature extraction from images refers to methods that aim at the effective description of the visual content of images. Many descriptors have been introduced for representing various image features and they can be divided in two main groups: hand-crafted and Deep Convolutional Neural Network (DCNN)-based descriptors. It should be noted that DCNN-based features outperform the hand-crafted features in most applications [5].

Hand-crafted features are divided into global and local descriptors. Global descriptors capture global characteristics of the image (e.g. the MPEG-7 [36] descriptor). Instead, local descriptors represent local salient points or regions and the most widely used are the SIFT descriptor [23] and its extensions (e.g. RGB-SIFT [33]), and the SURF descriptor [4] and its variations (e.g. CSURF [39]).

The most recent trend in feature extraction and image representation is learning features directly from the raw image pixels using DCNNs. These consist of many layers of feature extractors and can be used both as standalone classifiers, i.e., unlabeled images are passed through a pre-trained DCNN that performs the final class label prediction directly, or as generators of image features, i.e., the output of a hidden layer of the pre-trained DCNN is used as a global image representation [24, 37]. The latter type of features is referred to as DCNN-based and these features are used in the proposed framework due to their high performance in terms of both accuracy and efficiency.

Classification is the last step of the concept detection process. For learning the associations between the visual features and concept labels, algorithms such as Support Vector Machines (SVM) and Logistic Regression are used [24]. SVMs are trained separately for each concept, on ground-truth annotated corpora, and when a new unlabeled image arrives, the trained concept detectors return confidence scores that show the belief of each detector that the corresponding concept appears in the image.

### 5.5.1.2  Sky Detection Framework

In the employed framework, we train a 22-layer GoogLeNet [41] network on 5055 concepts, which are a subset of the 12,988 ImageNet concepts. Then, this network is applied on the TRECVID SIN 2013 development dataset and the output of the last fully-connected layer (5055 dimensions) is used as the input space of SVM classifiers trained on the 346 TRECVID SIN concepts. Among these classifiers, we use the one trained on the sky concept.

In order to evaluate the accuracy of the employed sky detection framework, we manually annotated (for the sky concept) 23,000 Instagram images (collected during preliminary past data collection activities) that were captured in the city of Berlin during the time period between 01/01/2016 and 15/04/2016. Sky detection was then applied on each image and the generated confidence scores were recorded in order to facilitate the selection of a decision threshold that provides a good trade-off between precision and recall. Based on this analysis, we opted for a 0.6 threshold (i.e. the sky concept is considered present if the confidence score is $\geq 0.6$) which led to 91.2% precision and 80.0% recall.

## 5.5.2 Sky Localization

Sky localization is an important computer vision problem which refers to the detection of all pixels that depict sky in an image. In this section, we first present the state of the art in sky localization (Sect. 5.5.2.1) and then describe the adopted sky localization approach which consists of the fusion of two diverse approaches, a deep learning-based one (Sect. 5.5.2.2) and one based on a set of heuristic rules (Sect. 5.5.2.3), that were found to work in a complementary manner (Sect. 5.5.2.4).

### 5.5.2.1 State of the Art

An approach that was proposed by Zhijie et al. [47] suggests measuring the sky border points. The authors propose several modifications of the original sky border position function, namely the determination of multi-border points for detecting complex sky regions in images. In [16], the authors suggest using blue color for localizing and tracking RGB color in different applications of image processing. Specifically, they propose a pixel based solution utilizing the sky color information. The success of deep networks on several domains led to their application in semantic segmentation as well. Specifically, several recent works have applied Convolutional Neural Networks (CNNs) to dense prediction problems, including semantic segmentation such as [8, 26, 29]; boundary prediction for electron microscopy by Ciresan et al. [6] and for natural images by a hybrid convnet/nearest neighbor model by Ganin and Lempitsky [9]. Moreover, Hariharan et al. [13] and Gupta et al. [12] adapt deep CNNs to semantic segmentation, but do so in hybrid detection-segmentation models. These approaches fine-tune a Regional-CNN system [11] by sampling bounding boxes and/or region proposals for detection, semantic segmentation, and instance segmentation. Finally, fully convolutional training is rare, but was used effectively by Tompson et al. [42] to learn an end-to-end part detector and spatial model for pose estimation.

### 5.5.2.2 FCN for Sky Localization

In the proposed framework, we employ the *fully convolutional network* (*FCN*) approach [22], which draws on recent successes of deep neural networks for image classification (e.g. [19]) and transfer learning. Transfer learning was first demonstrated on various visual recognition tasks (e.g. [7]), then on detection, and on both instance and semantic segmentation in hybrid proposal classifier models [11–13]. The work in [22] was the first to adapt deep classification architectures for image segmentation by using networks pre-trained for image classification and fine-tuned fully convolutionally on whole image inputs and per pixel ground truth labels. Importantly, it was shown [22] that the FCN approach achieves state-of-the-art segmentation performance in a number of standard benchmarks, including the SIFT Flow dataset where the FCN-16 variant achieved a pixel precision of 94.3% on the set of geometric classes, which include sky.

To measure the performance of the approach specifically on the task of sky localization, we used the SUN Database[19] [43], a comprehensive collection of annotated images covering a large variety of environmental scenes, places and the objects within. More specifically, we used the pre-trained (on the SIFT Flow dataset) FCN-16 model made available[20] by Long et al. [22], to predict the sky region of the 2030 SUN images for which the polygons capturing the sky part are provided. We measured a pixel precision of 91.77% and a pixel recall of 94.25%. It should be noted, that we are interested mainly in the precision of the approach given that what is required by the air quality estimation approach presented in Sect. 5.6 is recognizing accurately even a small part of the sky inside the image.

### 5.5.2.3 Sky Localization Using Heuristic Rules

The second approach for sky detection is based on heuristic rules that aim at recognizing the sky part of the images. The algorithm is based on identifying whether the pixels meet certain criteria involving their color values and the size of color clusters they belong to. The output of the algorithm is a mask containing all pixels that capture the sky. Figure 5.4 presents the pseudocode of the proposed method. It should be noted that the heuristic algorithm is far stricter than the FCN-based since sun and clouds are not considered part of the sky. Similarly to the FCN-based, the heuristic rule-based method was evaluated on the SUN database obtaining a mean precision of 82.45% and a mean recall of 59.22%.

---

[19]http://groups.csail.mit.edu/vision/SUN.

[20]https://github.com/shelhamer/fcn.berkeleyvision.org/tree/master/siftflow-fcn16s.

**Algorithm:** Sky localization using heuristic rule-based approach

        **Input:** Image (size MxN)

        **Output:** Image Mask

1.     Select upper 50% of the image
2.     **for** each pixel $p$ **do**
3.         **if** $(0.5 \leq \frac{R}{G} \leq 1) \wedge (0.5 \leq \frac{G}{B} \leq 1) \wedge (\frac{B}{R} > 1.25)$    is **false, then**
4.             $p$ is not part of the output mask
5.         **else**
6.             $p$ is candidate as part of the output mask
7.         **end if**
8.     **end for**
9.     **for** each $p$ **do**
10.        **if** there is at least one neighboring $p'$ that satisfies condition from step 3, **then**
11.            $p$ is candidate as part of the output mask
12.        **else**
13.            $p$ is not part of the output mask
14.        **end if**
15.     **end for**
16.    Finding Connected Components of remaining pixels
17.    **for** each pixel $p$ candidate of the output mask **do**
18.        **if** $p$ belongs to connected component with size over (MxN)/400 is **true, then**
19.            $p$ is part of the output mask
20.        **else**
21.            $p$ is not part of the output mask
22.        **end if**
23.     **end for**
24.    Initialize $Sp = 0$
25.    **for** each $p$ belonging to the output mask **do**
26.        $Sp = Sp + 1$
27.    **end for**
28.    **if** $Sp \geq \frac{MxN}{100}$   **then**
29.        Image does not contain significant part of clear sky
30.        **exit**
31.    **end if**
32.    **for** each $p$ belonging to the mask **do**
33.        **if** $\left(\frac{R}{G} > mean_R + 4 \cdot std_R\right) \,\&\&\, \left(\frac{R}{G} < mean_R - 4 \cdot std_R\right)$   is **true, then**
34.            $p$ is not part of the output mask
35.             **goto step 24**
36.        **end ifend forInitialize** $i = 0$
37.    **for** $i < 20$ **do**
38.        **if** G/B for pixels in vertical line $i$ increases monotonically is **false, then**
39.            Image does not contain significant part of sky and is discarded
40.            **exit**
41.        **end if**
42.    **end for**
43.    Image Mask produced from remaining pixels

**Fig. 5.4** Flowchart of the heuristic sky localization algorithm

### 5.5.2.4    Comparison of Sky Localization Methods

As already mentioned both localization methods were evaluated on the SUN database and the results showed that the *FCN* approach performed significantly better than the *heuristic* approach. However, a visual inspection of the ground truth annotations of the collection's images, revealed that the image region that is annotated as "sky" is not always suitable for air quality (AQ) estimation as in many cases the sky part is not clear (e.g. contains clouds, the sun, small objects). For these reasons, a specialized evaluation of the two sky localization methods that focuses explicitly on their ability to correctly identify sky regions that are suitable for AQ estimation is presented. To this end, out of about one million images that were collected with the Flickr and the webcam image collectors, we filtered out those in which the detection confidence of the sky concept is not very high (<0.8) to ensure that most of the remaining images will depict sky and then we took a random sample of 100 Flickr and 100 webcam images. For each image, sky masks were extracted using both approaches and the following questions were collaboratively answered by the authors:

- Q1-a: Does the image contain a sky region usable for AQ estimation? (Y/N)
- Q1-b: Please shortly describe the reason if you answered No to Q1-a.
- Q2: Is the sky region selected with the *FCN* approach usable for AQ estimation? (Y/N)
- Q3: Is the sky region selected with the *heuristic* approach usable for AQ estimation? (Y/N)

To ease the task, annotators were provided with horizontally aligned composite images where the masks generated by each approach were placed next to the original image (see Fig. 5.5).

The first question (Q1-a) aims at helping us identify images with a sky region usable for AQ estimation, so that we can subsequently evaluate the different sky localization methods only on images with a usable sky region. The responses to Q1-a revealed that both for Flickr and webcams images about 60% of the images contain a sky region that is usable for AQ estimation ("Yes" to Q1-a), while looking at the distribution of responses to Q1-b, we see that in most cases and for both types of images, it is the presence of clouds or cirrus clouds or the fact that the image is captured too early in the morning or too late in the evening that render images unusable for AQ estimation, despite the existence of a sky region.

Having identified images with usable sky regions, we focused on the ability of each sky localization approach to extract these regions. The results are presented in Table 5.1, which shows the percentages of correctly detected image regions using the *FCN* (Q2) and the *heuristic* (Q3) approach for Flickr and webcam images. At a first glance, the performance of the two methods appears much worse than the performance obtained on the SUN database. Note, however, that the evaluation performed here is much stricter as even if a small percentage of the region recognized as sky includes non-sky elements, then the whole region is marked incorrect. We observe that in contrast to the results obtained when the evaluation was performed on the SUN database, the *heuristic* approach performs

**Table 5.1** Percentages of correctly/incorrectly detected sky regions using each sky localization approach for Flickr and webcam images

| Method | Q2-*FCN* (Y/N) | Q3-*heuristic* (Y/N) |
|--------|----------------|----------------------|
| Flickr | 28.8%/71.2% | 45.8%/54.2% |
| webcams | 20.7%/79.3% | 50.0%/50.0% |

**Fig. 5.5** Comparison of the masks generated by the *FCN* approach (2nd column) with the masks generated by the *heuristic* approach (3rd column) for the images of the 1st column. The 4th column shows the masks generated by the *FCN+heuristic* approach

better than the *FCN* approach as it manages to correctly detect the sky region in 45.76%/50.00% of the Flickr/webcam images versus only 28.81%/20.69% for the *FCN* approach.

To better understand the merits of each approach, we performed a visual comparison of the generated masks (two examples are shown in Fig. 5.5). The comparison reveals that the masks generated by the *heuristic* approach are more-fine grained (e.g. small objects and text overlays that are common in webcam images are successfully filtered out), leading to more cases where all pixels identified as sky are actually sky ("Yes" to Q2) compared to the *FCN* approach ("Yes" to Q3). The *FCN* approach, on the other hand, is much better at avoiding "big" mistakes (e.g. recognizing sea, buildings or windows as sky). Motivated by the complementarity of the two approaches, we decided to develop a sky localization approach that combines them (*FCN+heuristic*). More specifically, we first calculate the sky mask using the *FCN* approach and then apply the *heuristic* algorithm, considering only those pixels that have been recognized as sky by the *FCN* approach. This way, we exploit the effectiveness of the *FCN* approach in roughly recognizing the sky region of the image and then utilize the *heuristic* approach to discard small non-sky elements. As can be seen in the right-most column of Fig. 5.5, *FCN+heuristic* generates much better sky masks than either of the two approaches alone.

Besides this qualitative evaluation, we also performed a quantitative evaluation of *FCN+heuristic*, as we did for the *FCN* and *heuristic* approaches, i.e. we collected responses to the question: "Q4: Is the sky region selected with the *FCN+heuristic* approach usable for AQ estimation? (Yes/No)" for the same set of 100 Flickr and

**Table 5.2**  Comparison of *FCN*, *heuristic* and *FCN+heuristic* sky localization approaches

|              | *FCN*                      | *heuristic*                | *FCN+heuristic* |
|--------------|----------------------------|----------------------------|-----------------|
| Accuracy     | 24.8%                      | 47.9%                      | **80.3%**       |
| Time/hardware| 103 ms/Nvidia GTX1070      | 125 ms/Intel i7-3770       | 128 ms          |

100 webcam images. The results of this evaluation are presented in Table 5.2, which shows the percentages of correctly, when considering all images (Flickr and webcams). As expected, there is a very large improvement as **80.34%** of the sky regions are correctly recognized by *FCN+heuristic*, compared to 47.86% for the *heuristic* approach and 24.79% for the *FCN* approach. Table 5.2 also reports the average (over 200 images) running time of the methods, when images are first downscaled to a maximum size of 250,000 pixels (respecting the aspect ratio). We see that both *FCN* and *heuristic* take slightly more than 100 ms per image on average, while *FCN+heuristic* has a running time that is only slightly higher, due to the fact that *heuristic* has to operate only on the pixels that are recognized as sky by *FCN*.

## 5.6  Air Quality Estimation Based on Sky Color Statistics

Aerosols are tiny particles suspended in the atmosphere which are emitted by natural as well as human activities (volcanoes, desert dust, forest fires, sea salt biomass burning, combustion of fossil fuel, industrial activities, etc.) [35]. Apart from impairing the quality of the air, they determine the levels of surface solar radiation by scattering and absorbing the light that comes from the sun [14]. Their scattering and absorbing efficiency depends on their macrophysical, microphysical and microchemical properties. So, aerosols, depending on their abundance and type, leave their mark on the radiation that reaches the ground.

A number of passive remote sensing instruments (e.g., sunphotometers, spectrophotometers) are capable of retrieving aerosol optical properties such as Aerosol Optical Depth (AOD) by measuring the radiation that reaches the ground at specific wavelengths. As the instruments originally measure light intensities in order to assign the measured light intensities to a specific AOD usually a Look-Up-Table (LUT) approach is followed. LUTs are produced with the use of a radiative transfer model (RTM). RTMs calculate the intensity of the light transferred within the atmosphere under different user-input scenarios that include information about the position of sun (solar zenith angle) relative to Earth and various atmospheric parameters (e.g., clouds, aerosols, water vapor, ozone, surface albedo, etc.). This way, one knows what the expected light intensity for specific atmospheric conditions is. By comparing these measured light intensities with those from a LUT, an estimate of the AOD can be retrieved.

According to the discussion above, the color (RGB) of the sky is expected to be determined partly by the amount and type of aerosols in the atmosphere. To date there have been some scientific efforts around the world to retrieve atmospheric aerosol properties from images taken from different types of digital cameras (e.g., [15, 27, 32]) and from paintings (e.g., [44, 45]). These efforts have returned promising results so far and further improvement is ongoing. The method followed in this work is based on the use of the ratio of the red and green band of the light (R/G) and the ratio of green and blue band of the light (G/B) from images. The main idea is that R/G and G/B depend on the amount and type of aerosols in the atmosphere [32, 44].

We decided to follow this method for a number of reasons. First of all, as discussed above, the method has already been validated in previous studies. It is based on the physics of light propagation through the atmospheric medium, contrary to approaches based on statistics or machine learning. This allows for a better understanding of the atmospheric processes that lead to high and low ratios in the images and makes it easier to understand the uncertainties and limitations of the method and proceed to corrections. In addition, the use of ratios instead of single-band RGB values compensates for biases emerging from factors such as the camera type, exposure time, sky viewing angle, etc. The LUT approach (see below) constitutes the basis of aerosol retrievals in atmospheric remote sensing, from ground-based instruments to satellite sensors. The same LUT could be used for retrieving the same quantities with images from passive remote sensing instruments in the future, allowing for a more direct validation of the method. Finally, the method is also fast, allowing its use on an operational basis.

The procedure that was followed for the production of the LUT is similar to the one described in [44] but more detailed as it takes into account the special characteristics of each region on a monthly basis, namely the optical properties of the aerosols such as the single scattering albedo and the asymmetry parameter, the ozone total column, the water vapor and the surface albedo. First, a LUT with the R/G and G/B was produced in order to assign R/G and G/B values to various aerosol loads. We use the aerosol optical depth at 550 nm ($AOD_{550}$) as a measure of the aerosol load in the atmosphere.

To produce the LUT we implemented radiative transfer simulations using the SBDART (Santa Barbara DISORT Atmospheric Radiative Transfer) radiative transfer model [31]. The radiative transfer equation is solved using the DISORT (Discrete Ordinate Radiative Transfer) method [40]. Sixteen streams were used. An IDL (Interactive Data Language) code that "feeds" SBDART with the necessary input data and executes the radiative transfer model for clear sky conditions was developed [2, 3]. The diffuse radiance (radiant flux received by a surface per unit solid angle per unit projected area) for the visible wavelength range (400–700 nm) was calculated. The diffuse radiance values at 700 nm (Red) were divided by the diffuse radiance values at 550 nm (Green) to get the R/G values and the radiance values at 550 nm (Green) were divided by the radiance values at 450 nm (Blue) to get the G/B values. Our tests showed that for specific sky viewing angles and azimuth angles (direction relative to the sun) in summer one should use G/B instead of R/G ratios as it is difficult to distinguish medium from high aerosol conditions with R/G ratios.

The globe was divided into 2592 grid cells with a resolution of $5° \times 5°$ and a sub-LUT was created for each cell in order to take into account the special characteristics of each region (optical properties of the aerosols, the ozone total column, the water vapor column and the surface albedo) (Fig. 5.6). The radiative transfer model was executed for clear-sky conditions for the central coordinates of each grid cell. This was done for various days within a year, times within a day, sky viewing angles, azimuth angles and for various $AOD_{550}$ bins, taking into account the special characteristics of each grid cell (input data). All these parameters are crucial for the radiative transfer calculations and taking into account their spatial and temporal variability increases the accuracy of the results. The core input data come from global climatologies and reanalysis projects. The aerosol optical properties (single scattering albedo and asymmetry parameter) come from the MACv1[21] (Max-Planck-Institute Aerosol Climatology version 1) climatology [18], the total ozone column, the water vapor column and surface albedo come from the ECMWF's ERA-interim reanalysis dataset[22] and the elevation data used in the calculations come from the U.S. Geological Survey GTOPO30 product.[23]

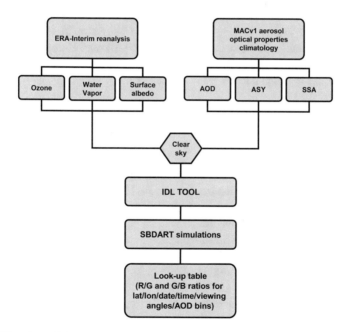

**Fig. 5.6** Flowchart of the method followed for the production of the LUT

---

The result of the radiative transfer calculations is a LUT consisting of 2592 ASCII columnar files (sub-LUTs), one for each $5° \times 5°$ grid cell. The sub-LUTs include the R/G ratio (where Red: 700 nm and Green: 550 nm), and the G/B ratio (where Green: 550 nm and Blue: 450 nm) for different days within a year, hours within a day and $AOD_{550}$ bins. After going through a number of tests for specific spots it was decided that sky viewing angles of 45° and azimuth angles of 90° should be used as the majority of user generated photos are close to this scenario.

The final step of the retrieval procedure includes the calculation of the $AOD_{550}$ that corresponds to the photo R/G ratio. This is done by calculating the difference of the LUT R/G ratio values that appear in the sub-LUT that corresponds to the geographical coordinates of the photo with the photo R/G values and selecting the $AOD_{550}$ value from the sub-LUT that minimizes this difference. As discussed above, only for summer and for specific sky viewing and azimuth angles G/B ratios are used instead of R/G ratios. Similarly to [45], the errors in $AOD_{550}$ should be less than 0.05 for values around 0.1 and can be up to 0.18 for $AOD_{550}$ values greater than 0.5. To avoid the uncertainties inserted in cases of large solar zenith angles the method is not applied to images taken close to the sunrise or sunset.

So far, the method has been tested for various places in Greece and in Europe. Results from three tests implemented for the city of Thessaloniki, Greece (an aerosol hot spot for the region of Eastern Mediterranean: [10]) and Europe as a whole are presented here, showing that the use of R/G (G/B) ratios is capable of revealing urban as well as regional particle pollution features.

## 5.6.1   Test 1

On 10/6/2016 from 18:10 to 18:55 (local timezone) we crossed Thessaloniki, Greece (1.5 million inhabitants) using the bus from one side of the city to the other following the coastline. A photo was taken each time the bus stopped in front of a bus stop (see Fig. 5.7 for the position of the 39 bus stops). The 39 photos were taken at a viewing angle of ~45° and an azimuth angle of ~30° relative to the sun. The photos were processed in order to calculate the R/G ratio. The results show that the R/G ratio increases gradually as one gets into the city centre. The R/G ratio decreases for an extended area covered with green and trees in the centre of the city, then increases again and finally decreases gradually as the bus leaves the city centre. As the distance covered by the bus is nearly 16 km and the R/G levels have a reasonable variability taking into account the expected emissions in the city (busy streets, parks, etc.), the method seems to be adequate to characterize the aerosol variability within an urban centre. According to these results the method is expected to have a spatial representativeness of 1–2 km.

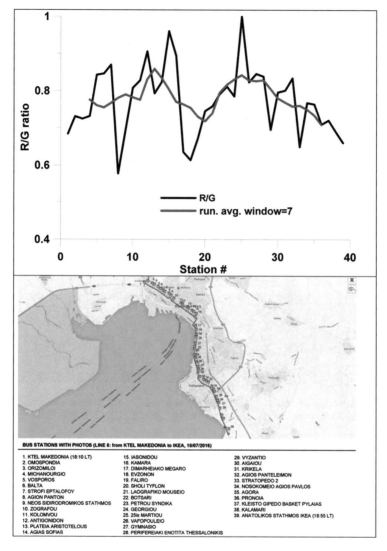

**Fig. 5.7** Bus stops where photos were taken during the Thessaloniki transect experiment (lower panel) and the corresponding R/G values for each one of the 39 stations (upper panel)

## 5.6.2  Test 2

Annual $AOD_{550}$ maps for Thessaloniki were produced using 435 Flickr images for the year 2012 (Fig. 5.8). Figure 5.8 was created using ordinary kriging for interpolation. The results were compared against $PM_{2.5}$ maps from [34] for Thessaloniki. In [34], the authors used a data assimilation algorithm coupling dispersion modeling and ground station data. The resulting $PM_{2.5}$ map of the metropolitan area

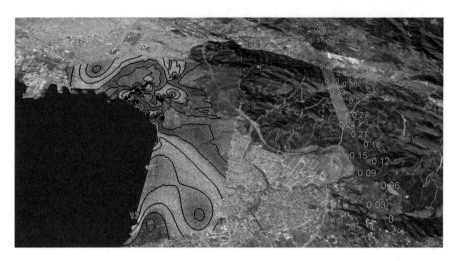

**Fig. 5.8** AOD levels over the city of Thessaloniki, Greece as retrieved from Flickr images for the year 2012

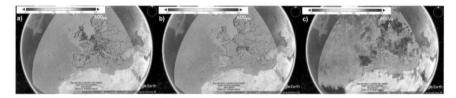

**Fig. 5.9** Comparison of results from Flickr (**a**) and webcam (**b**) images with MODIS/Terra satellite retrievals (**c**) from Europe. The data cover the period 24/2/2017 to 13/4/2017

of Thessaloniki reveals similar features with Fig. 5.7 (high pollution to the left of the port and pollution hot spots at the same locations) which adds further credibility to our results.

### 5.6.3 Test 3

We computed annual $AOD_{550}$ maps for Europe using ~31,000 Flickr and ~25,000 webcam images for the period March–April 2017 (Fig. 5.9a, b). Our results are compared against $AOD_{550}$ maps (Fig. 5.9c) with data from the MODIS/Terra satellite sensor (Col. 6, L3 data) which were acquired from NASA's Giovanni web database.[24] These maps show consistently high and low values over specific regions. All the maps share the same features (e.g., high values in N. Italy, Pays Bas, etc.). However, the Flickr images show better details than MODIS or webcam images. Hence, in the Flickr map several major cities are also seen.

---

[24]https://giovanni.gsfc.nasa.gov.

## 5.7 Conclusions and Future Work

The proposed framework comprises all the steps required for estimating air quality from publicly available images. The sources used for image retrieval are social media platforms and webcams. As far as the social media platforms that could be used, a study was realized that covered all the popular platforms that are used for image sharing. The results of the study revealed that Flickr is the most appropriate candidate due to the strict limitations on data usage imposed by the other social media platforms as well as the considerable amount of data uploaded to Flickr. A set of experiments regarding the images returned by Flickr covering Europe shows that the average number of geotagged images collected daily are approximately 5000. As far as webcams are concerned, two very large repositories of webcam images were analyzed, AMOS and webcams.travel. Both repositories were found to contain a significant number of webcams and, at the same time, offer a relatively simple way of retrieving images and other required information (location and time) from them. Consequently, two specialized collectors were implemented, facilitating the collection of images from approximately 3500 different European locations at regular time intervals.

All the collected images are processed using a three-step procedure. The first step involves sky detection, the second sky localization, and the third air quality estimation. Sky localization involves detecting the sky part of the image and two methods were studied. One based on Fully Convolutional Networks and one based on heuristic rules proposed by air quality experts. An evaluation of the two techniques was realized, showing that the two methods achieve better results when applied in a complementary way. Eventually, for the sky part of the images the R/G and G/B ratios are calculated and air quality estimation is realized. A number of atmospheric aerosol measurements using personal photos, images from Flickr and from webcams for the city of Thessaloniki, Greece and Europe was produced to study the ability of the method to reveal local and regional pollution features. The first comparisons with results from previous studies and with satellite observations highlight the potential of the method.

The evaluation of the proposed framework showed that results are promising. However, there is still room for improvement with respect to the accuracy of the sky detection and localization methods and the spatial and temporal resolution of the LUTs. It has been shown that the presence of cirrus clouds is in many cases the reason why an image is considered unsuitable for air quality estimation. Even though in many cases it is difficult to decide whether an image is unsuitable for air quality estimation due to the presence of cirrus clouds even with a naked eye, a possible direction for future work would be the development of a specialized concept detector that would automatically recognize and filter sky-depicting images where sky is covered by this type of clouds or the use of haze as proposed in other works for estimating air quality.

As a final remark, we would like to point out that the very promising results of the proposed framework as well as results of a number of other recent works on image-based air quality estimation, on one hand highlight the potential of using images as cheap air quality sensors but on the other hand highlight the importance of evaluating all these approaches under a common evaluation framework in order to draw more reliable conclusions with respect to their relative merits. The development of such a benchmark is a promising direction for future work.

**Acknowledgements** This work is partially funded by the European Commission under the contract number H2020-688363 hackAIR.

# References

1. Aiello, L.M., Petkos, G., Martin, C., Corney, D., Papadopoulos, S., Skraba, R., Göker, A., Kompatsiaris, I., Jaimes, A.: Sensing trending topics in twitter. IEEE Transactions on Multimedia **15**(6), 1268–1282 (2013)
2. Alexandri, G., Georgoulias, A., Meleti, C., Balis, D., Kourtidis, K., Sanchez-Lorenzo, A., Trentmann, J., Zanis, P.: A high resolution satellite view of surface solar radiation over the climatically sensitive region of eastern mediterranean. Atmospheric Research **188**, 107–121 (2017)
3. Alexandri, G., Georgoulias, A., Zanis, P., Katragkou, E., Tsikerdekis, A., Kourtidis, K., Meleti, C.: On the ability of regcm4 regional climate model to simulate surface solar radiation patterns over europe: an assessment using satellite-based observations. Atmospheric Chemistry and Physics **15**(22), 13,195–13,216 (2015)
4. Bay, H., Ess, A., Tuytelaars, T., Gool, L.V.: Speeded-up robust features (surf). Computer Vision and Image Understanding **110**(3), 346–359 (2008). Similarity Matching in Computer Vision and Multimedia
5. Chatfield, K., Simonyan, K., Vedaldi, A., Zisserman, A.: Return of the devil in the details: Delving deep into convolutional nets. arXiv preprint arXiv:1405.3531 (2014)
6. Ciresan, D., Giusti, A., Gambardella, L.M., Schmidhuber, J.: Deep neural networks segment neuronal membranes in electron microscopy images. In: Advances in neural information processing systems, pp. 2843–2851 (2012)
7. Donahue, J., Jia, Y., Vinyals, O., Hoffman, J., Zhang, N., Tzeng, E., Darrell, T.: Decaf: A deep convolutional activation feature for generic visual recognition. In: International conference on machine learning, pp. 647–655 (2014)
8. Farabet, C., Couprie, C., Najman, L., LeCun, Y.: Learning hierarchical features for scene labeling. IEEE transactions on pattern analysis and machine intelligence **35**(8), 1915–1929 (2013)
9. Ganin, Y., Lempitsky, V.: Nˆ 4-fields: Neural network nearest neighbor fields for image transforms. In: Asian Conference on Computer Vision, pp. 536–551. Springer (2014)
10. Georgoulias, A.K., Alexandri, G., Kourtidis, K.A., Lelieveld, J., Zanis, P., Pöschl, U., Levy, R., Amiridis, V., Marinou, E., Tsikerdekis, A.: Spatiotemporal variability and contribution of different aerosol types to the aerosol optical depth over the eastern mediterranean. Atmospheric Chemistry and Physics **16**(21), 13,853 (2016)
11. Girshick, R., Donahue, J., Darrell, T., Malik, J.: Rich feature hierarchies for accurate object detection and semantic segmentation. In: Proceedings of the IEEE conference on computer vision and pattern recognition, pp. 580–587 (2014)
12. Gupta, S., Girshick, R., Arbeláez, P., Malik, J.: Learning rich features from rgb-d images for object detection and segmentation. In: European Conference on Computer Vision, pp. 345–360. Springer (2014)

13. Hariharan, B., Arbeláez, P., Girshick, R., Malik, J.: Simultaneous detection and segmentation. In: European Conference on Computer Vision, pp. 297–312. Springer (2014)
14. Haywood, J., Boucher, O.: Estimates of the direct and indirect radiative forcing due to tropospheric aerosols: A review. Reviews of geophysics **38**(4), 513–543 (2000)
15. Igoe, D., Parisi, A., Carter, B.: Characterization of a smartphone camera's response to ultraviolet a radiation. Photochemistry and photobiology **89**(1), 215–218 (2013)
16. Irfanullah, K.H., Sattar, Q., Sadaqat-ur Rehman, A.A.: An efficient approach for sky detection. IJCSI International Journal of Computer Science Issues **10** (2013)
17. Jacobs, N., Roman, N., Pless, R.: Consistent temporal variations in many outdoor scenes. In: Computer Vision and Pattern Recognition, 2007. CVPR'07. IEEE Conference on, pp. 1–6. IEEE (2007)
18. Kinne, S., O'Donnel, D., Stier, P., Kloster, S., Zhang, K., Schmidt, H., Rast, S., Giorgetta, M., Eck, T.F., Stevens, B.: Mac-v1: A new global aerosol climatology for climate studies. Journal of Advances in Modeling Earth Systems **5**(4), 704–740 (2013)
19. Krizhevsky, A., Sutskever, I., Hinton, G.E.: Imagenet classification with deep convolutional neural networks. In: Advances in neural information processing systems, pp. 1097–1105 (2012)
20. Li, Y., Huang, J., Luo, J.: Using user generated online photos to estimate and monitor air pollution in major cities. In: Proceedings of the 7th International Conference on Internet Multimedia Computing and Service, p. 79. ACM (2015)
21. Liu, C., Tsow, F., Zou, Y., Tao, N.: Particle pollution estimation based on image analysis. PloS one **11**(2), e0145,955 (2016)
22. Long, J., Shelhamer, E., Darrell, T.: Fully convolutional networks for semantic segmentation. In: Proceedings of the IEEE Conference on Computer Vision and Pattern Recognition, pp. 3431–3440 (2015)
23. Lowe, D.G.: Distinctive image features from scale-invariant keypoints. International journal of computer vision **60**(2), 91–110 (2004)
24. Markatopoulou, F., Mezaris, V., Patras, I.: Cascade of classifiers based on binary, non-binary and deep convolutional network descriptors for video concept detection. In: Image Processing (ICIP), 2015 IEEE International Conference on, pp. 1786–1790. IEEE (2015)
25. Moumtzidou, A., Papadopoulos, S., Vrochidis, S., Kompatsiaris, I., Kourtidis, K., Hloupis, G., Stavrakas, I., Papachristopoulou, K., Keratidis, C.: Towards air quality estimation using collected multimodal environmental data. In: International Workshop on the Internet for Financial Collective Awareness and Intelligence, pp. 147–156. Springer (2016)
26. Ning, F., Delhomme, D., LeCun, Y., Piano, F., Bottou, L., Barbano, P.E.: Toward automatic phenotyping of developing embryos from videos. IEEE Transactions on Image Processing **14**(9), 1360–1371 (2005)
27. Olmo, F.J., Cazorla, A., Alados-Arboledas, L., López-Álvarez, M.A., Hernández-Andrés, J., Romero, J.: Retrieval of the optical depth using an all-sky ccd camera. Applied optics **47**(34), H182–H189 (2008)
28. Pan, Z., Yu, H., Miao, C., Leung, C.: Crowdsensing air quality with camera-enabled mobile devices. In: AAAI, pp. 4728–4733 (2017)
29. Pinheiro, P., Collobert, R.: Recurrent convolutional neural networks for scene labeling. In: International Conference on Machine Learning, pp. 82–90 (2014)
30. Poduri, S., Nimkar, A., Sukhatme, G.S.: Visibility monitoring using mobile phones. Annual Report: Center for Embedded Networked Sensing pp. 125–127 (2010)
31. Ricchiazzi, P., Yang, S., Gautier, C., Sowle, D.: Sbdart: A research and teaching software tool for plane-parallel radiative transfer in the earth's atmosphere. Bulletin of the American Meteorological Society **79**(10), 2101–2114 (1998)
32. Saito, M., Iwabuchi, H.: A new method of measuring aerosol optical properties from digital twilight photographs. Atmospheric Measurement Techniques **8**(10), 4295–4311 (2015)

33. van de Sande, K.E.A., Gevers, T., Snoek, C.G.M.: Evaluating color descriptors for object and scene recognition. IEEE Transactions on Pattern Analysis and Machine Intelligence **32**(9), 1582–1596 (2010)
34. Sarigiannis, D.A., Karakitsios, S.P., Kermenidou, M.V.: Health impact and monetary cost of exposure to particulate matter emitted from biomass burning in large cities. Science of The Total Environment **524**, 319–330 (2015)
35. Seinfeld, J.H., Pandis, S.N.: Atmospheric chemistry and physics: from air pollution to climate change. John Wiley & Sons (2016)
36. Sikora, T.: The mpeg-7 visual standard for content description-an overview. IEEE Transactions on Circuits and Systems for Video Technology **11**(6), 696–702 (2001). 10.1109/76.927422
37. Simonyan, K., Zisserman, A.: Very deep convolutional networks for large-scale image recognition. arXiv preprint arXiv:1409.1556 (2014)
38. Snoek, C., Cappallo, S., Fontijne, D., Julian, D., Koelma, D.C., Mettes, P., van de Sande, K., Sarah, A., Stokman, H., Towal, R., et al.: Qualcomm research and university of amsterdam at trecvid 2015: Recognizing concepts, objects, and events in video. In: NIST TRECVID Workshop (2015)
39. Spyromitros-Xioufis, E., Papadopoulos, S., Kompatsiaris, I., Tsoumakas, G., Vlahavas, I.: A comprehensive study over vlad and product quantization in large-scale image retrieval. IEEE Transactions on Multimedia (2014)
40. Stamnes, K., Tsay, S.C., Wiscombe, W., Jayaweera, K.: Numerically stable algorithm for discrete-ordinate-method radiative transfer in multiple scattering and emitting layered media. Applied optics **27**(12), 2502–2509 (1988)
41. Szegedy, C., Liu, W., Jia, Y., Sermanet, P., Reed, S., Anguelov, D., Erhan, D., Vanhoucke, V., Rabinovich, A.: Going deeper with convolutions. In: Proceedings of the IEEE conference on computer vision and pattern recognition, pp. 1–9 (2015)
42. Tompson, J.J., Jain, A., LeCun, Y., Bregler, C.: Joint training of a convolutional network and a graphical model for human pose estimation. In: Advances in neural information processing systems, pp. 1799–1807 (2014)
43. Xiao, J., Hays, J., Ehinger, K.A., Oliva, A., Torralba, A.: Sun database: Large-scale scene recognition from abbey to zoo. In: Computer vision and pattern recognition (CVPR), 2010 IEEE conference on, pp. 3485–3492. IEEE (2010)
44. Zerefos, C., Gerogiannis, V., Balis, D., Zerefos, S., Kazantzidis, A.: Atmospheric effects of volcanic eruptions as seen by famous artists and depicted in their paintings. Atmospheric Chemistry and Physics **7**(15), 4027–4042 (2007)
45. Zerefos, C., Tetsis, P., Kazantzidis, A., Amiridis, V., Zerefos, S., Luterbacher, J., Eleftheratos, K., Gerasopoulos, E., Kazadzis, S., Papayannis, A.: Further evidence of important environmental information content in red-to-green ratios as depicted in paintings by great masters. Atmospheric Chemistry and Physics **14**(6), 2987–3015 (2014)
46. Zhang, C., Yan, J., Li, C., Rui, X., Liu, L., Bie, R.: On estimating air pollution from photos using convolutional neural network. In: Proceedings of the 2016 ACM on Multimedia Conference, pp. 297–301. ACM (2016)
47. Zhijie, Z., Qian, W., Huadong, S., Xuesong, J., Qin, T., Xiaoying, S.: A novel sky region detection algorithm based on border points. International Journal of Signal Processing, Image Processing and Pattern Recognition **8**(3), 281–290 (2015)

# Chapter 6
# Traits: Structuring Species Information for Discoverability, Navigation and Identification

**Thomas Vattakaven, Prabhakar Rajagopal, Balasubramanian Dhandapani, Pierre Grard, and Thomas Le Bourgeois**

**Abstract** Conventionally, species traits concepts have been conceived from an ecological perspective after grouping them as functional traits, response traits or effect traits: attributes of individual organisms that express phenotypes in response to the environment and its effects on the organism. From an informatics perspective, traits may be conceived to encompass a broader vocabulary that can capture any species attribute including, but not limited to those concerning its morphology, taxonomy, functional role, habitat, ecological interactions, trophic strategies, genetics, evolution, conservation status, anthropological uses, ecosystem services etc. The evolution of such a vocabulary and its standardisation across disciplines and taxa is a challenge, but one that needs imminent attention as the field develops. Furthermore, traits can have values that vary within and across individuals and species. The ability to associate traits with levels of a taxonomic hierarchy, aggregate species traits from individual records, flexibility to attribute categorical text, numeric, temporal and spatial values; associate them with ontologies; and conform to standards, can evolve traits as a flexible framework to structure descriptive, numeric and tabular data on species. Such a framework for structuring descriptive species data will, allow better discoverability and navigation of the information and has potential for developing

T. Vattakaven (✉) · P. Rajagopal
Strand Life Sciences, Bangalore, India

B. Dhandapani
French Institute of Pondicherry, UMIFRE 21 CNRS-MAEE, Pondicherry, India
e-mail: balu.d@ifpindia.org

P. Grard
CIRAD, Nairobi, Kenya
e-mail: pierre.grard@cirad.fr

T. Le Bourgeois
CIRAD, UMR AMAP, Montpellier, France
e-mail: thomas.le_bourgeois@cirad.fr

© Springer International Publishing AG, part of Springer Nature 2018
A. Joly et al. (eds.), *Multimedia Tools and Applications for Environmental & Biodiversity Informatics*, Multimedia Systems and Applications,
https://doi.org/10.1007/978-3-319-76445-0_6

further applications such as polyclave identification keys and analytical aids for big data. The open source Biodiversity Informatics Platform that powers three international initiatives across Asia and Africa has been evolving as an effective platform to aggregate and build open access databases for varied biodiversity data types. It has ability to handle varied data types such as descriptive data, occurrences, maps and documents. The platform has recently added a traits infrastructure that is participatory and can aggregate traits from curated databases as well as by crowdsourcing from observation and collection data. It is flexible in building vocabularies to structure descriptive species information and media, evolving into a framework which allows flexible yet efficient navigation of species information in an information system. Here, we discuss this model, its application within the applied initiatives, its potential use in classifying multimedia data for species characterization in a complex context and in facilitating trait analysis. We also cover potential applications of the trait framework for developing into a comprehensive and effective infrastructure for aggregating and structuring species information.

## 6.1  Introduction

Species diversity has long been held to be a representation of the health of an ecosystem. However, there has been a paradigm shift around this approach and current opinion is that the health of an ecosystem may be dictated more by the diversity of traits within its species and the functionality they provide; not merely the species diversity [36, 50, 51]. Traits are increasingly being used in conservation planning and policy making such as in protecting parrotfish in Belize for their traits helping coral reef health [13] or classifying forests in Peru using canopy traits to rebalance its conservation portfolio [6].

The traditional concept of species traits is thought to have originated from attempts to classify organisms based on their role and function and has since evolved to be an integral part of modern functional ecology [24, 42, 53]. However, a straightforward definition of traits has not been defined and the trait concept continues to evolve and adapt to suit the needs of ecological study. Traits have been variously defined as *"the fundamental descriptors of organismal phenotype, functionality and performance"* [5] or *"a well-defined, measurable property of organisms, usually measured at the individual level and used comparatively across species"* [40] and *"morphological, biochemical, physiological, structural, phenological, or behavioural characteristics of organisms that influence how they respond to the environment and/or their effects on ecosystem properties"* [53]. Most definitions view traits from the point of its function (functional traits) further classifying them either based on their effect on ecosystems (effect traits) or based on their response to environmental conditions affecting the ecosystem (response traits). Although the traits are now a part of scientific terminology, traits as a concept is more generic and finds usage within the common language, across multiple disciplines [53].

## 6.2   Our Implementation of the Traits Infrastructure

The India Biodiversity Portal (IBP[1]) is an open access biodiversity information system for India launched in December 2008 [52]. IBP's two main objectives are (a) aggregating curated biodiversity data for all species in India and (b) nurturing a community where biodiversity amateurs and experts can interact. The portal is participatory and all information is freely and openly accessible by any member of the public under Creative Commons licences. The IBP platform code base is open source and is licensed under GNU General Public License. The same platform and its code base is used to power the Bhutan Biodiversity Portal (BBP)[2] and the Weed Identification and Knowledge in the Western Indian Ocean (WIKWIO)[3] [35] portal.

The platform consists of five interconnected modules. Species Pages cater to aggregating descriptive content on species. The Observation module facilitates crowd sourcing of observations of species and curation of the information through citizen science. The Maps module aggregates spatial data layers on a variety of themes related to biodiversity and the Documents module allows gathering of information from academic journals and grey literature on species. In some of the portals an identification module that utilises the IDAO identikit system [34], primarily based on morphological species traits is also functioning. All modules are interconnected and the Species pages function as the aggregating end point where all species information is summarised and displayed.

Over the years, each portal has been aggregating significant descriptive content within species pages. These are structured under species fields contained under concepts, categories and subcategories adapted from standard controlled vocabulary such as the Species Profile Model and Plinian Core. These categories include varied subjects such as taxonomy, natural history, habitat and distribution of species. The IBP alone contains over 26,000 descriptive pages which have textual descriptions pertaining to the biology of a species and provide end users with rich information. However, as more information is aggregated, searching, retrieving and navigating it becomes difficult. Searching and querying textual content returns matches that are often unsatisfactory or doesn't fulfil the context of the query. Although information may be present within textual descriptions, it becomes onerous to find references to specific trait categories such as particular functional roles, phenotypic characters, conservation classifications etc. of an organism from within the large descriptive text. Structuring of species characters enable easier and more efficient querying of species data and provide the basis for a host of different applications that can take advantage of it to create intelligent solutions that are discussed later in this chapter. We have designed and set up an infrastructure that allows a trait based structuring of

---

[1] http://indiabiodiversity.org.
[2] http://biodiversity.bt/.
[3] http://portal.wikwio.org/.

species information on the open source Biodiversity Informatics Platform (BIP[4]). Our implementation of this first version of the traits infrastructure is described below.

The trait infrastructure allows the creation of a **Trait**. Traits in this context encompasses a broader vocabulary than its conventional definition in ecology, that can capture any species attribute including, but not limited to those concerning its morphology, taxonomy, functional role, habitat, ecological interactions, trophic strategies, genetics, evolution, conservation status, anthropological uses, ecosystem services, geographical location etc. Each trait has a string value as its **trait_name** and can be associated with a value datatype of either string, date, numeric, boolean or color. The **trait_type** decides if the trait can hold single (single categorical) eg: herb; or multiple (multiple categorical) values (eg: yellow, red, green) or a range (eg: January–March) from possible categorical variables. Numeric traits are associated with a measurement **unit** (eg: centimeter, millimeter). Each trait can be represented with an **icon** and a **description** (or an illustrated glossary on the portal) and the possibility of associating it with a term in a defined **ontology** on the web. Attribution for the trait can be provided in the **source field** (Fig. 6.1).

All traits are considered to be species characters; however, some traits may vary between individuals of the same species, depending on species polymorphism, ecological conditions, or observer perception. These traits where variables can be collected from observations of each individual can be marked as **show_in_observation**. Such traits will be considered as observation traits and will display as input fields for observers when an observation is uploaded. These observation traits capture the individual variation of the trait within a species. Observation traits require further summarization and aggregation of its collective values for defining a species character. In addition, certain traits such as the endangered status (IUCN Redlist[5]) can be marked as **is_not_observation_trait** as it is never an observed value of a species and is flagged as such. Finally, observation traits can further be set as **is_participatory** or not. Observation traits that are participatory can be edited by any user by examining media on the observation eg: the phenological state of a plant or the sex of a bird. Traits that are not participatory can only be recorded by the original observer in the field and during the observation eg: the abundance of the species in the locality.

Each trait is associated with a taxon via its **trait_taxonomy_definition**. This defines the taxonomic scope of the trait. The taxon rank may be at any single level between the Kingdom and Infraspecies or multiple taxa at the same level eg: multiple families of butterflies. It may be left undefined to associate it as a "root trait" applicable to all taxa. Further, a trait is associated with a species field via its **field_id**. Species fields are contained under concepts, categories and subcategories adapted from standard controlled vocabulary such as the Species Profile Model and Plinian Core eg: Habitat and Distribution > Distribution > Description. The species

---

[4]https://github.com/strandls/biodiv.

[5]http://www.iucnredlist.org/.

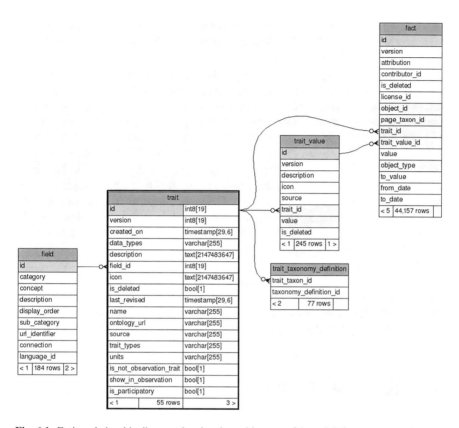

**Fig. 6.1** Entity relationship diagram showing the architecture of the trait infrastructure on the open source biodiversity informatics platform

field provides the context to the structured vocabulary and allows it to be covered by existing exchange standards like the Darwin Core for species descriptions. The species fields are also utilised for grouping traits for display within species pages and observation templates.

Traits are always associated with a set of **values** whose properties are inherited from the trait's defined *data_type* and *trait_type*. These may be strings, numbers, colors or numeric values. Values also have the ability to store a value name, icons, descriptions and attribution text.

**Facts** qualify the value(s) of a trait per species and is associated with a taxon eg: The value of the trait "Growth form" for the species *Mangifera indica* is 'tree'. A fact can be defined at either the observation level or species level. Facts can be stored with an attribution, a contributor and a licence. This allows the fact source to be traceable and facilitates unambiguous attribution for sharing and data exchange. A fact is always a combination of the trait and its value and can be comprised of a string, date or date range, numeric or numeric range, a boolean value or color. The first version of the traits infrastructure has adopted a flat, non-hierarchical structure for defining traits, eliminating issues with traits and values in turn being part of a hierarchy themselves. Plant leaf shape:acuminate instead of Plant > Plant part > Leaf > Leaf shape > acuminate.

In this system, every species page is provided with all the traits that have been associated to its taxon via the taxonomic scope of the traits (Fig. 6.2). New traits and their values can be created via an online interface. Species pages and their traits are editable by experts who have been authorised to input species information. The expert can annotate species pages with trait values by choosing one or more values depending on the *trait_type*. There are also interfaces for uploading large data-sheets containing traits and values as columns and rows per species. When traits have been tagged as *show_in_observation*, the traits will be available for annotation by any contributor who uploads an observation. Observation forms (Fig. 6.3) require the observer to categorize the observation as a mammal, bird, fish, amphibian, reptile, molluscs, arthropod, plant or fungus. Depending on the defined taxonomic scope of a trait, relevant traits are displayed for annotation by the observer on choosing the category.

The option of making annotatable traits available within observations facilitates crowd-sourcing of trait values for any trait. Such crowd-sourced trait values in combination with temporal and spatial information can then be aggregated to derive species trait states. Eg: Phenological states of a tree in an area (Fig. 6.4).

## 6.3 Applications of a Trait Based Infrastructure

### 6.3.1 Structuring Information

Historically, textual write-up with illustrations has served the purpose of describing species accounts. Descriptive text, supplemented by visual media in the form of photographs, diagrams or other multimedia types have been used to convey the morphology, biology, ecology and use of species, generating tomes of textual content for many species. Most of this content is generated as human language easily understandable to human beings who have a comprehension of the specific vocabulary. It is expected that the large volumes of data that is emerging in biodiversity sciences will pave the way for it to join the other "big data" science subjects such as astronomy and particle-physics [49]. However, most of this content is not structured, making it difficult for software agents (softbots) to process this information. The pathway for information extraction (IE) from existing biodiversity information into structured formats can be a complicated process, involving digitisation, optical character recognition (OCR), employing established IE templates for automated Named Entity Recognition (NER), verification, curation and ensuring organisation of the output in standards compliant structured formats. Thus, much of the potential of this rich information remains untapped. With the advent of the information age, easy availability of processing power and with automation rapidly gaining momentum, there is a need to structure human language for processing by machines.

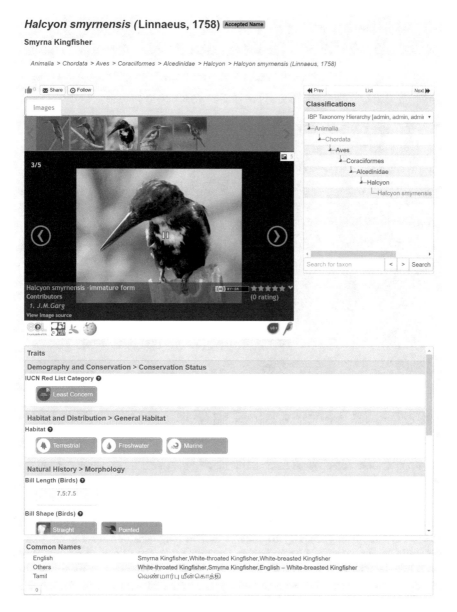

**Fig. 6.2** A screenshot of a species page of *Halcyon smyrnensis* (Linnaeus 1758) on the portal showing the traits annotated for the species

Natural Language Processing (NLP) is an emerging area where computers are programmed to analyze, understand and process human language in a manner that can organize and structure. NLP may be utilised for performing tasks such

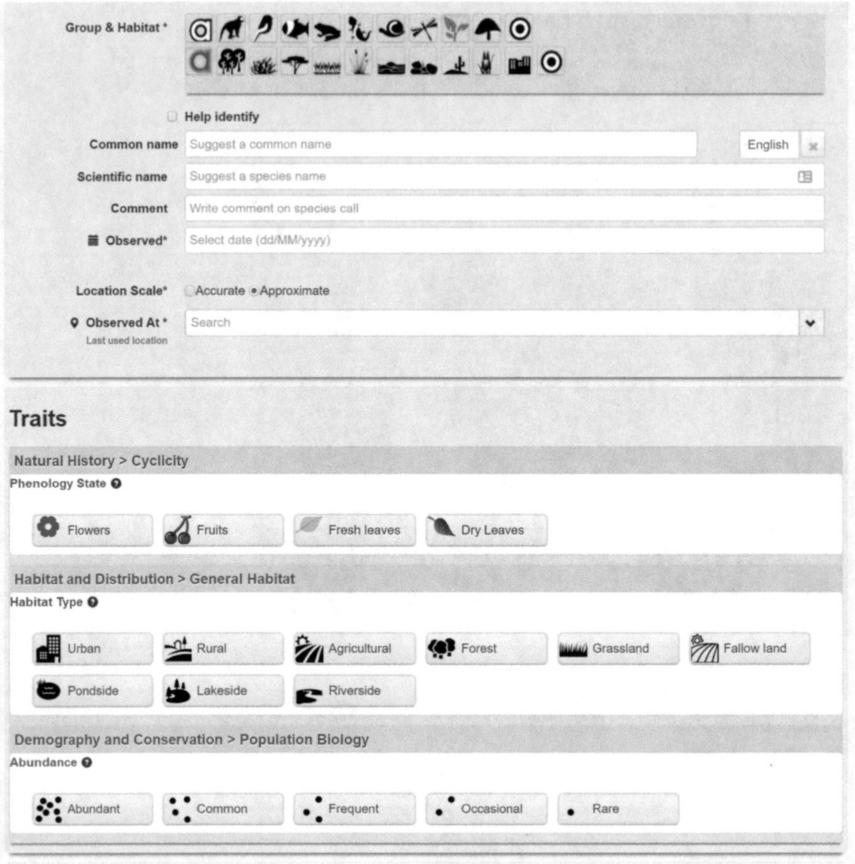

**Fig. 6.3** The Observation upload form on IBP showing traits available for annotation within an observation during upload

summarising content, extracting relationships, categorising named entities and more. For this, it is necessary to convert human language into a more structured semantic language that can be understood by both machines as well as people. However, human language is verbose and understanding it requires an understanding of not only the words, but also the concepts associated with words. Ambiguities in natural language syntax and reference makes natural language processing and understanding a very difficult area in computer science. However, the science of NLP has made considerable progress of late and there is much choice in the stack of techniques that can be utilised, depending upon the level of sophistication required to interpret language. Typically, NLP employs syntactic parsing and "chunking" to identify and tag parts of speech within a sentence through the help of "Taggers". Readers may refer to Thessen [49] for a deeper review of various parsers and taggers available and the complexity of tasks that they can perform. An example of NLP

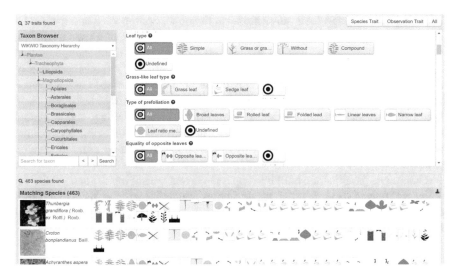

**Fig. 6.4** A screenshot of the trait filter page on the Biodiversity Informatics Platform allowing querying of all species in the system by trait constraints

applied in the biodiversity context include the NetiNeti project, a machine learning based approach for recognition of scientific names including the discovery of new species names from text [2].

An alternative approach is in structuring text along controlled vocabularies and ontologies which facilitate expression and easy understanding while enabling machine access and computability. The concept of a Semantic Web [9] that marks up all web content to specified ontologies seeks to facilitate this, but has fallen short of its objective due to difficulties in formulating ontologies for numerous subjects and finding semantic mappings between different ontologies [18]. There are currently hundreds of controlled vocabularies but none are comprehensive enough to be used as a reliable source for training machine learning systems [15]. However, when such structured vocabularies exist, either as dictionaries, glossaries, gazetteers, or ontologies, they can serve as knowledge entities that can drive further NLP systems [49]. Readers are pointed towards a review by Cui [16], on the techniques that have been used for automated annotation, some exploratory results on characteristics of morphological descriptions challenges facing automated annotation systems.

The approach we have implemented on the biodiversity portal platform, is a simple approach, designed to facilitate an opportunistic structuring of species information through a controlled, yet flexible vocabulary that can be annotated through crowdsourcing mechanisms. The IBP has implemented a system where any user may request rights to edit species content within species pages. On verification of the user's expertise and allocation of rights, they are able to add descriptive text, upload multimedia content, validate records or structure species pages with traits and values. IBP currently has hundreds of such expert contributors who have been enriching species pages. All transactions on the portal are recorded and listed on a user profile page. IBP has been developing the fundamentals of a user reputation system, which captures, classifies and calculates both categorized and aggregated activity scores for each user. Such reputation systems have shown promise in motivating and rewarding user participation in crowd-sourced and citizen science web portals [47]. They can also contribute towards automatically identifying users with expertise and help in allocating them rights for contributing and validating content. The availability of annotatable traits in a structured, objective type format within species pages, provides further incentive for participating users to develop content.

Games and gamification built into apps are currently a hot topic. Such games can entice volunteers to accomplish tasks such as annotating species pages with trait data. Although the BIP has not utilised this mode yet, it may be well worth exploring to drive user participation as noted in game-based citizen science initiatives such as Questa[6] and Happy Match.[7]

Over time, with enhanced contribution and participation by experts and observation providers, it is expected that all species pages will accumulate structured content with traits and values categorised by species fields and their ontology.

### 6.3.2   Bridging Species with Non-taxonomical Linkages

The most common way of classifying species is through the modern biological classification based on the evolutionary relationships between organisms i.e taxonomic or systematic classification. All organisms within a common taxonomic rank usually share the same characteristics. Originally these were morphological characters at the organismal level but has progressively shifted to the cellular and the molecular level. In taxonomic classification systems homologous character states may be captured in the closest common ancestral rank of these species. However, analogous character states, functional characters, classifications based on niche occupation, categorizations based ecosystem services or conservation values etc. that may cut across taxonomic ranks cannot be captured with taxonomic categorization alone. A classification infrastructure that is flexible enough to capture various classification categories, cutting across taxonomic ranks and categories, while still retaining linkages with a taxonomic classification is required. Trait-based approaches towards categorising species can help establish cross-linkages across species as well as establish its role within a biogeographical niche, to help model species interactions, dispersal ability, and physiological responses of organisms to changes in biogeographical conditions [54].

---

[6]https://questagame.com/.

[7]http://www.citizensort.org/web.php/happymatch.

The current trait infrastructure that IBP has built, allows non-taxonomical linkages by allowing categorization of species into traits and values for any aspect of a species character. This will allow grouping of all species by a specific habitat or niche eg: marine intertidal zone; or conservation status eg: Critically endangered; or Flower color: Yellow; and other traits. While biodiversity information systems continue to gather data on species diversity and richness, it is acknowledged that there is great shortfall in functional richness data in the form of traits. Cross-linking species through traits will be valuable means of adding a dimension on the functional diversity and facilitate upscaling of analysis to larger populations allowing us to predict the consequences of global changes for ecosystem functions and services [25].

### 6.3.3   Using Traits for Species Identification

The flexibility of the trait infrastructure in its ability to connect species across non-taxonomic linkages does not preclude its usage in taxonomic scenarios. The trait infrastructure may be utilised to create flexible identification systems that can overcome the limitations of dichotomous keys ie. two choices at each branching point. Such keys have evolved from its usage within traditional taxonomic literature in the form of handwritten content and printed media.

Species identifications are key to non-specialised enthusiasts who include photographers, naturalists and hobbyists; as well as more serious ecology researchers as most ecological studies require identification of species [19]. It is also required for farmers and agricultural extensionists who need to identify weeds at an early stage of growth for which classical flora based on flower structure can not work, for better weed management [33]. However, most keys are set up as polyclave keys that offer a fixed sequence of identification steps with each step having only two alternatives. Such keys are not well suited to both non-specialised and specialised users due to its requirement of starting from the top level and having enough information on the organism to work one's way down the key without skipping any character and to arrive at an identification. It is also complicated, as most keys use technical jargon which a casual user seeking to identify a species or species group may not understand [39]. Furthermore, it assumes that a user has all the information available to proceed from the top question to the subsequent levels. This makes it impossible to proceed down the key when limited information in the form of a flower or leaf only is available.

An alternative to dichotomous keys is the Polyclave keys or Random/Multi access keys where a unit represents the state of character for a species and the user is allowed to input the state of the character for his unidentified specimen in any order [38]. This facilitates the user to enter the most obvious and observable character states and eliminate species not having those states, enabling him to further narrow the species identification by looking for and providing more details [19]. If morphological characters of species were held as traits, and their character states as values, the trait infrastructure would function well as polyclave/random access identification key that will work well without the need for further modification. On selecting any trait value of a trait, the system is able to eliminate species lacking the selected value and output a list of possible matching species.

One of the first steps in discovering and understanding biodiversity is to identify organisms. Indeed the Convention on Biodiversity has explicitly made calls for increasing numbers of para-taxonomists to facilitate identification of species for ecological studies. Graphical multi-access identification systems provided by the trait-based system will facilitate easier identification of species even for non-specialised users and help work towards enabling practitioners in parataxonomy. The traits infrastructure implemented on IBP allows the common non-specialist user and field ecologists to navigate in any order and sequence of the characters observed to filter, narrow searches and identify species. This in addition to the platform's ability to harness crowd-sourced participation in identification and validation of content.

### 6.3.4 Querying by Traits

It is well known that there are definite links between a species' biological traits and its environment [48]. Such relationships have been studied and elucidated over centuries of research and extrapolation. With modern information technology, it may be possible to run simple queries to extract hidden relationships and patterns between a species and its environment. The benefit of having a system that can categorise species non-taxonomically, taxonomically and additionally associate spatial and temporal characteristic of species is that it can serve as a mechanism to combine queries that address different types of categorisation. Eg: Endangered species of the family *Liliopsida* having compound leaves and a fruiting period in August. Such combinations of traits, though important, will be constricted by the availability of the data as trait values. Such query ability would be of great value in functional ecology, where functional traits can be queried with combinations of effect and response traits along with constraints from other categorizations to elucidate patterns and trends. Combinations of species traits and environmental parameters are already being explored as a mechanism to predict responses to issue such as climate change and estimate its 'range-shift capacity' [20]. Such queries would be valuable in the case of conservation and planning for prioritising species of concern and also for management of exotic invasive species.

### 6.3.5 Traits for Organizing Multimedia on Species

Databases that aggregate multimedia on species such as observational reports of organisms often accumulate huge amounts of multimedia. Such media objects are often associated with spatial, temporal and taxonomic data. However there is tremendous scope to further categorise such media based on descriptors that annotate specific events or content captured within the media. Traits can serve as a content-based descriptor for categorizing and organising multimedia.

Our design of the trait infrastructure allows associating traits with observational records containing media. Annotating observations with traits allows the aggregation of variable trait values for a species. Media objects associated with an observation can inherit the annotated trait value as a categorization. Eg: An observation of the Common Mormon butterfly (*Papilio polytes*) annotated with the observation traits Life stage: adult and Sex: male will likely contain media objects of a male, adult *Papilio polytes* species and can be categorised as such by inheriting the traits. Observations are also associated with a species which in turn have non-variable traits which can be inherited directly for categorizing the media objects. Eg: The Common Mormon butterfly may hold species level traits for Predominant wing color: black, Body color: Black and Wing-spot color: white. Such information can be inherited directly from the species to categorize observation multimedia, i.e media of a butterfly with black wings, black body and white spots.

Multimedia categorization is intended to create stable multimedia representations, analysis, processing and inference schemes that can in turn be used for indexing, querying, retrieving, summarising and associating multimedia objects within different usage contexts [3]. It is valuable in the context of automating media for displaying within species page fields as illustrations of a character, or habit. It could be used within graphical identification systems to guide users towards a species identification. Categorised multimedia can be used for training machine learning algorithms to run further automated categorizations, image recognition and identification.

## 6.4   Summary and Scope for Expansion

### 6.4.1   Traits and Ontology

In the recent decades, the rate at which biological information is being generated has increased dramatically due to advances in technology such as with embedded and autonomous sensor networks, remote-sensing platforms, and long-term monitoring projects [7]. In response to the large data generated, there has been a proliferation of databases aimed at facilitating access to it. This is also true of trait data. Most of the larger trait databases are specialised in on some aspect of biodiversity—The TRY[8] specialises in plant trait data [31]. Biolflor[9] is another large database on plant traits.

---

[8]https://www.try-db.org/TryWeb/Home.php.

[9]http://www2.ufz.de/biolflor/index.jsp.

SeaLifeBase[10] is a trait database that specialises in traits of multicellular marine organisms. PanTHERIA aggregates data on a set of key life-history, ecological and geographical traits of all known extant and recently extinct mammals[11] [30]. However, aggregating data from different databases and sharing data between them is not seamless as different databases have terminological, syntactic, and semantic variations that need to be overcome to equate interrelated data [27].

Ontologies are crucial towards achieving this. An ontology is defined as "a formal specification of a shared conceptualization" [11] and is a controlled vocabulary that describes objects and the relations between them in a formal way so that data represented by it becomes more accessible to both people and software agents [27]. Some of the prominent ontology services that are relevant for trait data include the Phenotypic Quality Ontology (PATO) [37], the Uber Anatomy Ontology (UBERON) [41], Plant Trait Ontology (PO) [28], Vertebrate Trait Ontology [45](Park et al. 2013), Environments Ontology (ENVO) [12] etc. The Ontobee [43] is a linked data server which dereferences and presents individual ontology term URIs, aimed to facilitate ontology data sharing, visualization, query, integration, and analysis. Flora Phenotype Ontology (FLOPO) is an example of an ontology that has been extracted from existing ontologies using automated reasoning [26]. Ontologies provide the leeway to construct flexible vocabularies while still maintaining linkages with standards. Hierarchical ontologies also provide a crucial role in providing relationships and context among the concepts. The hierarchical ontology classifies the concepts at each level and proceeds from generalized to specialized concepts usually through a **"is_a"** relation. However, there are practical difficulties in incorporating ontologies for reasons such as ambiguity in trait name meanings, inability to match hierarchical positions, multiplicity in terminologies and difficulty in placing concepts under parent terms [32]. In a traits framework, it is difficult to adhere to a defined ontology as traits vary with and between taxonomic groups. Current ontologies are also not comprehensive enough to be used as a reliable source for training machine learning systems [15]. Our current framework has left provisions for incorporating an ontology url for every trait but does not insist on it. The ontology needs in our framework will be revisited and evolved as better standards evolve. Our approach is similar to that of TraitBank. TraitBank is a searchable, comprehensive, open digital repository for organism traits built into the framework of the Encyclopedia of Life project. It mobilizes data from diverse sources and simply links data records to relevant ontologies and controlled vocabularies [46]. The meaning of each attribute is analysed manually and a formally-defined semantic terms are used to represent them. Trait bank has chosen to create provisional URIs for TraitBank terms that are not yet a part of the most relevant ontologies and its vocabulary is not yet standardized. TraitBank[12] data is downloadable as well as accessible through its search API under Creative Commons

---

[10]http://www.sealifebase.org/.

[11]https://ecologicaldata.org/wiki/pantheria.

[12]http://eol.org/info/516.

licence. TraitBank uses and extends TDWG Darwin Core standards which is the most widely used standard for exchange of biodiversity data and hence is reusable by other initiatives including our BIP platform.

### 6.4.2  Traits for Machine Learning and Image Analysis

Machine learning is programming computers to optimize a performance criterion using example data or past experience [4]. As a part of this process a machine (i.e., computer algorithm) improves its performance automatically with experience [55]. The learning may occur in various ways which may include rule sets, decision trees, clustering algorithms, linear models, Bayesian networks, artificial neural networks, and custom algorithms [49]. Machine learning algorithms can be trained with pre-classified data and setting determinants that will result in predictions as outcomes for new data (supervised machine learning). It can be used in pattern recognition, data mining, inference and estimations, signal processing etc.

Recent advances in computer vision have facilitated the development of algorithms that help classify plants based on images of their leaves. The LeafSnap[13] project is an excellent example of this. The Pl@ntNet[14] project is another example of machine learning employed in identifying plants from images, supplemented by validated observations to improve its recognition performance with the increasing data [10, 29]. Recently, large biodiversity based citizen science initiatives such as iNaturalist and Cornell Lab of Ornithology through its Merlin Bird ID[15] app have implemented image recognition of species to suggest identifications. Both of these rely on the Visipedia project[16] [8]—a network of people and machines that is designed to harvest and organize visual information and make it openly accessible. Visipedia in turn uses the TensorFlow open-source software library for numerical computation using data flow graphs, developed by Google [1].

In the context of the open source biodiversity informatics platform that aggregates voluminous multimedia content and classifies it using a traits infrastructure, there is great potential for the data to be employed in training machine learning algorithms for several workflows. On a primary level the system requires that users uploading observations choose a broad categorisation for the organism being uploaded. The IBP currently has over 82,730 images of plants, 57,902 of arthropods, 28,029 of birds, 4797 of reptiles, 4535 of mammals 2665 fungi, 2611 of amphibians, 912 fishes, and 591 of molluscs. These images are continuously being incremented and can be used to train machine learning algorithms that can offer suggestions to classify observations into these primary classes when

---

[13]http://leafsnap.com/.

[14]http://identify.plantnet-project.org/.

[15]http://merlin.allaboutbirds.org/.

[16]https://sites.google.com/visipedia.org/index/home.

a user uploads images. On a secondary level, within the above categorization, most observations are identified to at least a family level taxon. Each family and many genus have hundreds to thousands of images which again can be employed in algorithms that can offer suggestions for the taxonomic grouping that the observation belongs to. Finally, finer morphological level traits are associated with the species. More complicated workflows which are able to distinguish an organism within an image, either automatically or through user input, can utilise trait based information to offer suggestions on a species identification. Eg: predominant + associated colors of a bird's plumage + beak shape; or Leaf shape + leaf type + flower color. On the other end of the spectrum, traits not yet recorded or established can be extracted from multimedia images of an identified species. Eg: IBP has hundreds of images of the lizard species identified as *Calotes versicolor*. Based on the color patterns and overall body/head color extracted from image analysis, it may be possible to append values for these traits as color values to the species.

Within its species pages, the different BIP portals have rich descriptive text, classified broadly along taxonomy and morphology, natural history, habitats and distribution, uses and management, conservation etc. By its nature, biological descriptions are highly specialised, has great diversity in content as well as syntactic variation across taxa [49]. However, there have been several initiatives that have been able to extract morphological traits from morphological descriptions using combinations of keyword matching, contextual pattern matching and parsing techniques. The Worldwide Botanical Knowledge Base,[17] the Terminator project [17] and the MARTT biosemantics project[18] [14] are examples of some initiatives that have experimented with this to varying degrees [49]. In future, the BIP portal may be able draw from such learning to automate the extraction of traits from descriptive text. Even without machine extraction, the trait infrastructure can be employed to crowdsource values such as flower colors or seed size. It is important to capture perspectives of experts as well as of casual observers. For example leaves of *Bidens pilosa* can be considered simple but deeply lobed by a botanist or as a compound leaf by a non botanist. In other cases, the apex of the leaf can be round, cuneate or acuminate for different individuals of the same species and depending on ecological conditions where the plant is growing (shade/sun, dry/humid). These combinations of different user's observation of a species trait and variations in the plant's polymorphism if captured effectively can help to drive identification systems such as the IDAO system [23] with added intelligence that can be used and enriched by users of varied expertise.

As more open data becomes available in biodiversity sciences, the idea of creating a biodiversity knowledge graph [44], where all biodiversity data is a network of connected entities, such as taxa, taxonomic names, publications, people, species, sequences, images, and collections is moving towards realisation. However,

---

[17]http://wwbota.free.fr/.

[18]https://sites.google.com/site/biosemanticsproject/project-progress-wiki.

there are still large data gaps in our knowledge of biodiversity [21]. Much needs to be done on one hand to liberate data acquired in legacy formats over the past 250 years and at the same time keep pace with data and formats emerging in the modern day that may contain valuable biological data. This is an ambitious goal that requires the combined efforts of both taxonomy and technology. Trait data forms a vital part of the knowledge graph for interconnecting biodiversity data as well as being one of the Essential Biodiversity Variables (EBVs) that are required for developing indicators of global biodiversity change [22]. As we have discussed above, traits hold much potential along varied fronts—in driving data extraction, to providing an axis for data navigation and in constructing intelligent knowledge systems in biodiversity data and we expect to see a plethora of trait based applications emerging in the near future.

# References

1. Abadi, M., Agarwal, A., Barham, P., Brevdo, E., Chen, Z., Citro, C., Zheng, X. (2016). TensorFlow: Large-Scale Machine Learning on Heterogeneous Distributed Systems. Retrieved from http://arxiv.org/abs/1603.04467
2. Akella, L., Norton, C. N., & Miller, H. (2012). NetiNeti: discovery of scientific names from text using machine learning methods. BMC Bioinformatics, 13(1), 211. https://doi.org/10.1186/1471-2105-13-211
3. Aldershoff, F., Salden, A. H., Iacob, S. M., & Kempen, M. (2003). Supervised multimedia categorization. In M. M. Yeung, R. W. Lienhart, & C.-S. Li (Eds.) (p. 100). https://doi.org/10.1117/12.476242
4. Alpaydin, E. (2010). Introduction to machine learning. MIT Press.
5. Arnold, S. J. (1983). Morphology, Performance and Fitness. American Zoologist, 23(2), 347–361. http://doi.org/10.1093/icb/23.2.347
6. Asner, G. P., Martin, R. E., Knapp, D. E., Tupayachi, R., Anderson, C. B., Sinca, F., Llactayo, W. (2017). Airborne laser-guided imaging spectroscopy to map forest trait diversity and guide conservation. Science, 355(6323). Retrieved from http://science.sciencemag.org/content/355/6323/385
7. Baker, K. S., & Millerand, F. (2010). Infrastructuring ecology: challenges in achieving data sharing Karen S. Baker and Florence Millerand. In Collaboration in the new life sciences / edited by John N. Parker, Niki Vermeulen and Bart Penders. Farnham, Surrey, England: Ashgate.
8. Barry, J. (2016). Identifying biodiversity using citizen science and computer vision: Introducing Visipedia. TDWG 2016 ANNUAL CONFERENCE. Retrieved from https://mbgocs.mobot.org/index.php/tdwg/tdwg2016/paper/view/1112/0
9. Berners-Lee, T., Hendler, J., & Lassila, O. (2001). The Semantic Web A new form of Web content that is meaningful to computers will unleash a revolution of new possibilities. Retrieved from https://pdfs.semanticscholar.org/566c/1c6bd366b4c9e07fc37eb372771690d5ba31.pdf
10. Bonnet, P., Joly, A., Goëau, H., Champ, J., Vignau, C., Molino, J.-F., Boujemaa, N. (2016). Plant identification: man vs. machine. Multimedia Tools and Applications, 75(3), 1647–1665. https://doi.org/10.1007/s11042-015-2607-4
11. Borst, W. N. (1997, September 5). Construction of Engineering Ontologies for Knowledge Sharing and Reuse. Centre for Telematics and Information Technology University of Twente University of Twente. Retrieved from https://research.utwente.nl/en/publications/construction-of-engineering-ontologies-for-knowledge-sharing-and-

12. Buttigieg, P., Morrison, N., Smith, B., Mungall, C. J., & Lewis, S. E. (2013). The environment ontology: contextualising biological and biomedical entities. Journal of Biomedical Semantics, 4(1), 43. https://doi.org/10.1186/2041-1480-4-43

13. Cernansky, R. (2017). Biodiversity moves beyond counting species. Nature, 546(7656), 22–24. http://doi.org/10.1038/546022a

14. Cui, H. (2008). Converting Taxonomic Descriptions to New Digital Formats. Biodiversity Informatics, 5(0). https://doi.org/10.17161/bi.v5i0.46

15. Cui, H. (2010a). Competency evaluation of plant character ontologies against domain literature. Journal of the American Society for Information Science and Technology, 61(6), n/a-n/a. https://doi.org/10.1002/asi.21325

16. Cui, H. (2010b). Semantic annotation of morphological descriptions: an overall strategy. BMC Bioinformatics, 11, 278. https://doi.org/10.1186/1471-2105-11-278

17. Diederich, J., Fortuner, R., & Milton, J. (1999). Computer-assisted data extraction from the taxonomical literature. Retrieved November 27, 2017, from https://www.math.ucdavis.edu/~milton/genisys/terminator.html

18. Doan, A., Madhavan, J., Domingos, P., & Halevy, A. (2004). Ontology Matching: A Machine Learning Approach. In Handbook on Ontologies (pp. 385–403). Berlin, Heidelberg: Springer Berlin Heidelberg.

19. Edwards, M., & Morse, D. R. (1995). The potential for computer-aided identification in biodiversity research. Trends in Ecology & Evolution, 10(4), 153–158. https://doi.org/10.1016/S0169-5347(00)89026-6

20. Estrada, A., Morales-Castilla, I., Caplat, P., & Early, R. (2016). Usefulness of Species Traits in Predicting Range Shifts. Trends in Ecology & Evolution, 31(3), 190–203. https://doi.org/10.1016/j.tree.2015.12.014

21. Geijzendorffer, I. R., Regan, E. C., Pereira, H. M., Brotons, L., Brummitt, N., Gavish, Y., Walters, M. (2016). Bridging the gap between biodiversity data and policy reporting needs: An Essential Biodiversity Variables perspective. Journal of Applied Ecology, 53(5), 1341–1350. https://doi.org/10.1111/1365-2664.12417

22. Guralnick, R. (2017). Traits as Essential Biodiversity Variables. Proceedings of TDWG, 1, e20295. https://doi.org/10.3897/tdwgproceedings.1.20295

23. Grard, P., Bonnet, P., Prosperi, M.-J., Le Bourgeois, T., Edelin, C., Theveny, F., & Alain, C. (2009). A graphical tool for computer-assisted plant identification. In Proceedings of TDWG 2009 Annual Conference. Montpellier, France. Retrieved from http://www.tdwg.org/proceedings/article/view/485

24. Hébert, M.-P., Beisner, B. E., & Maranger, R. (2017). Linking zooplankton communities to ecosystem functioning: toward an effect-trait framework. Journal of Plankton Research, 39(1), 3–12. http://doi.org/10.1093/plankt/fbw068

25. Hillebrand, H., & Matthiessen, B. (2009). Biodiversity in a complex world: consolidation and progress in functional biodiversity research. Ecology Letters, 12(12), 1405–1419. https://doi.org/10.1111/j.1461-0248.2009.01388.x

26. Hoehndorf, R., Alshahrani, M., Gkoutos, G. V., Gosline, G., Groom, Q., Hamann, T., Weiland, C. (2016). The flora phenotype ontology (FLOPO): tool for integrating morphological traits and phenotypes of vascular plants. Journal of Biomedical Semantics, 7(1), 65. https://doi.org/10.1186/s13326-016-0107-8

27. Hughes, L. M., Bao, J., Hu, Z.-L., Honavar, V., & Reecy, J. M. (2008). Animal trait ontology: The importance and usefulness of a unified trait vocabulary for animal species. Journal of Animal Science, 86(6), 1485–91. https://doi.org/10.2527/jas.2008-0930

28. Jaiswal, P., Ware, D., Ni, J., Chang, K., Zhao, W., Schmidt, S., McCouch, S. (2002). Gramene: Development and Integration of Trait and Gene Ontologies for Rice. Comparative and Functional Genomics, 3(2), 132–136. https://doi.org/10.1002/cfg.156

29. Joly, A., Bonnet, P., Goëau, H., Barbe, J., Selmi, S., Champ, J., Barthélémy, D. (2016). A look inside the Pl@ntNet experience. Multimedia Systems, 22(6), 751–766. https://doi.org/10.1007/s00530-015-0462-9

30. Jones, K. E., Bielby, J., Cardillo, M., Fritz, S. A., O'Dell, J., Orme, C. D. L., Purvis, A. (2009). PanTHERIA: a species-level database of life history, ecology, and geography of extant and recently extinct mammals. Ecology, 90(9), 2648–2648. https://doi.org/10.1890/08-1494.1
31. Kattge, J., Daz, S., Lavorel, S., Prentice, I. C., Leadley, P., Bönisch, G., Wirth, C. (2011). TRY - a global database of plant traits. Global Change Biology, 17(9), 2905–2935. https://doi.org/10.1111/j.1365-2486.2011.02451.x
32. Khan, S., & Safyan, M. (2014). Semantic matching in hierarchical ontologies. Journal of King Saud University - Computer and Information Sciences, 26(3), 247–257. https://doi.org/10.1016/j.jksuci.2014.03.010
33. Le Bourgeois, T., E. Jeuffrault, P. Grard and A. Carrara (2004). A new process to identify the weeds of La Réunion Island: the AdvenRun system. 14th Australian Weeds Conference, Charles Sturt University, Wagga Wagga, Australia, Weed Society of New South Wales.
34. Le Bourgeois, T., P. Bonnet, M. Couteau, P. Grard, C. Edelin, J. Prosperi and F. theveny (2008). IDAO Identification assisted by computer. IUCN World Conservation Congress. Workshop : Safegarding biodiversity and livelyhood from biological invasion: goal sharing of experience and information as a key step to effective management at local level. Barcelona, Spain.
35. Le Bourgeois, T., P. Grard, A. P. Andrianaivo, A. Gaungoo, Y. Ibrahim, J. A. Randria-mampianina, D. Balasubramanian, P. Marnotte, B. Ramesh, V. Andrianavalona, F. Hadji, Y. Karthik, M. Ramamonjihasina, K. Sathish and A. Seechurn (2015). WIKWIO - Weed Identification and Knowledge in the Western Indian Ocean - Web 2.0 participatory portal., European Union programme ACP S&T II, Cirad, IFP, MCIA/MSIRI, FOFIFA, CNDRS eds.http://portal.wikwio.org.
36. Loreau, M., Naeem, S., Inchausti, P., Bengtsson, J., Grime, J. P., Hector, A., Wardle, D. A. (2001). Biodiversity and Ecosystem Functioning: Current Knowledge and Future Challenges. Science, 294(5543). Retrieved from http://science.sciencemag.org/content/294/5543/804
37. Mabee, P., Ashburner, M., Cronk, Q., Gkoutos, G., Haendel, M., Segerdell, E., Westerfield, M. (2007). Phenotype ontologies: the bridge between genomics and evolution. Trends in Ecology & Evolution, 22(7), 345–350. https://doi.org/10.1016/j.tree.2007.03.013
38. Mata-Montero, E., & Carranza-Rojas, J. (2016). Automated Plant Species Identification: Challenges and Opportunities (pp. 26–36). Springer, Cham. https://doi.org/10.1007/978-3-319-44447-5_3
39. Martellos, S., & Nimis, P. L. (n.d.). KeyToNature: Teaching and Learning Biodiversity: Dryades, the Italian Experience.
40. McGill, B. J., Enquist, B. J., Weiher, E., & Westoby, M. (2006). Rebuilding community ecology from functional traits. Trends in Ecology & Evolution, 21(4), 178–185. http://doi.org/10.1016/j.tree.2006.02.002
41. Mungall, C. J., Torniai, C., Gkoutos, G. V, Lewis, S. E., & Haendel, M. A. (2012). Uberon, an integrative multi-species anatomy ontology. Genome Biology, 13(1), R5. https://doi.org/10.1186/gb-2012-13-1-r5
42. Nock, C. A., Vogt, R. J., Beisner, B. E., Nock, C. A., Vogt, R. J., & Beisner, B. E. (2016). Functional Traits. In eLS (pp. 1–8). Chichester, UK: John Wiley & Sons, Ltd. https://doi.org/10.1002/9780470015902.a0026282
43. Ong, E., Xiang, Z., Zhao, B., Liu, Y., Lin, Y., Zheng, J., He, Y. (2017). Ontobee: A linked ontology data server to support ontology term dereferencing, linkage, query and integration. Nucleic Acids Research, 45(D1), D347–D352. https://doi.org/10.1093/nar/gkw918
44. Page, R. (2016). Towards a biodiversity knowledge graph. Research Ideas and Outcomes, 2, e8767. https://doi.org/10.3897/rio.2.e8767
45. Park, C. A., Bello, S. M., Smith, C. L., Hu, Z.-L., Munzenmaier, D. H., Nigam, R., Reecy, J. M. (2013). The Vertebrate Trait Ontology: a controlled vocabulary for the annotation of trait data across species. Journal of Biomedical Semantics, 4(1), 13. https://doi.org/10.1186/2041-1480-4-13
46. Parr, C. S., Schulz, K. S., Hammock, J., Wilson, N., Leary, P., Rice, J., & Corrigan, R. J. (2016). TraitBank: Practical semantics for organism attribute data. Semantic Web, 7(6), 577–588. https://doi.org/10.3233/SW-150190

47. Silvertown, J., Harvey, M., Greenwood, R., Dodd, M., Rosewell, J., Rebelo, T., McConway, K. (2015). Crowdsourcing the identification of organisms: A case-study of iSpot. ZooKeys, 480, 125–146. https://doi.org/10.3897/zookeys.480.8803
48. Statzner, B., Hildrew, A. G., & Resh, V. H. (2001). Species traits and environmental constraints: Entomological research and the history of ecological theory. Annual Review of Entomology, 46(1), 291–316. https://doi.org/10.1146/annurev.ento.46.1.291
49. Thessen, A. E., Cui, H., & Mozzherin, D. (2012). Applications of natural language processing in biodiversity science. Advances in Bioinformatics, 2012, 391574. https://doi.org/10.1155/2012/391574
50. Tilman, D., & Downing, J. A. (1994). Biodiversity and stability in grasslands. Nature, 367(6461), 363–365. http://doi.org/10.1038/367363a0
51. Tilman, D., Knops, J., Wedin, D., Reich, P., Ritchie, M., & Siemann, E. (1997). The Influence of Functional Diversity and Composition on Ecosystem Processes. Science, 277(5330). Retrieved from http://science.sciencemag.org/content/277/5330/1300
52. Vattakaven, T., George, R., Balasubramanian, D., Réjou-Méchain, M., Muthusankar, G., Ramesh, B., & Prabhakar, R. (2016). India Biodiversity Portal: An integrated, interactive and participatory biodiversity informatics platform. Biodiversity Data Journal, 4, e10279. http://doi.org/10.3897/BDJ.4.e10279
53. Violle, C., Navas, M.-L., Vile, D., Kazakou, E., Fortunel, C., Hummel, I., & Garnier, E. (2007). Let the concept of trait be functional! Oikos, 116(5), 882–892. http://doi.org/10.1111/j.0030-1299.2007.15559.x
54. Violle, C., Reich, P. B., Pacala, S. W., Enquist, B. J., & Kattge, J. (2014). The emergence and promise of functional biogeography. Proceedings of the National Academy of Sciences of the United States of America, 111(38), 13690–6. https://doi.org/10.1073/pnas.1415442111
55. Witten, I. H. (Ian H., Frank, E., & Hall, M. A. (Mark A. (2011). Data mining : practical machine learning tools and techniques. Morgan Kaufmann. Retrieved from http://www.sciencedirect.com/science/book/9780123748560

# Chapter 7
# Unsupervised Bioacoustic Segmentation by Hierarchical Dirichlet Process Hidden Markov Model

**Vincent Roger, Marius Bartcus, Faicel Chamroukhi, and Hervé Glotin**

**Abstract**  Bioacoustics is powerful for monitoring biodiversity. We investigate in this paper automatic segmentation model for real-world bioacoustic scenes in order to infer hidden states referred as song units. Nevertheless, the number of these acoustic units is often unknown, unlike in human speech recognition. Hence, we propose a bioacoustic segmentation based on the Hierarchical Dirichlet Process (HDP-HMM), a Bayesian non-parametric (BNP) model to tackle this challenging problem. Hence, we focus our approach on unsupervised learning from bioacoustic sequences. It consists in simultaneously finding the structure of hidden song units, and automatically infers the unknown number of the hidden states. We investigate two real bioacoustic scenes: whale, and multi-species birds songs. We learn the models using Markov-Chain Monte Carlo (MCMC) sampling techniques on Mel Frequency Cepstral Coefficients (MFCC). Our results, scored by bioacoustic expert, show that the model generates correct song unit segmentation. This study demonstrates new insights for unsupervised analysis of complex soundscapes and illustrates their potential of chunking non-human animal signals into structured units. This can yield to new representations of the calls of a target species, but also to the structuration of inter-species calls. It gives to experts a tractable approach for efficient bioacoustic research as requested in Kershenbaum et al. (Biol Rev 91(1):13–52, 2016).

## 7.1 Introduction

Acoustic communication is common in the animal world where individuals communicate with sequences of some different acoustic elements [3]. An accurate analysis is important in order to give a better identification of some animal species and

V. Roger (✉) · M. Bartcus · H. Glotin
DYNI Team, DYNI, Aix Marseille Univ, Université de Toulon, CNRS, LIS, Marseille, France

F. Chamroukhi
LMNO UMR CNRS, Statistics and Data Science, University of Caen, Caen, France

© Springer International Publishing AG, part of Springer Nature 2018
A. Joly et al. (eds.), *Multimedia Tools and Applications for Environmental & Biodiversity Informatics*, Multimedia Systems and Applications,
https://doi.org/10.1007/978-3-319-76445-0_7

113

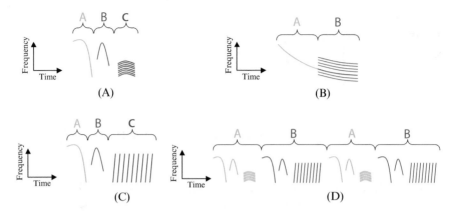

**Fig. 7.1** The four acoustic common ways used to divide into units [3]. (**a**) Separated by silence. (**b**) Change in acoustic properties (regardless of silence). (**c**) Series of sounds. (**d**) Higher levels of organisation

interpret the identified song units in time. There is a lack of methodologies focused on real world data, and with further applications in ecology and wildlife management. One of the major bottlenecks for the application of these methodologies is their inability to work under heavy complex acoustic environment, where different taxa may sing together or conversely, their extreme sensitivity which may result in an over classification due to the high degree of variability insight many repertoire of the vocal species. In this paper, we model the sequence of a non-human signals and determine their acoustic song units. The way according to which non-human acoustic sequences can be interpreted can be summarized as shown in Fig. 7.1. Four common properties are used to define potential criteria for segmenting such signals into song units. The first way, shown in Fig. 7.1a, consists in separating the signals using silent gaps. The second way, shown in Fig. 7.1b, consists in separating the signals according to the changes in the acoustic properties in the signal. The third way, shown in Fig. 7.1c consists in grouping similar sounds separated with silent gaps as a single unit. The last common way, shown in Fig. 7.1d consists in separating signal in organized sound structure, considered as fundamental units.

Manual segmentation is time consuming and not possible for a large acoustic dataset. That is why automatic approaches are needed. Furthermore, in bioacoustic signals, the problem of segmenting signals of many species, is still an issue. Hence, a well-principled learning system based on unsupervised approach can help to have a better understanding of bioacoustics species. In this context, we investigate statistical latent data models to automatically identify song units. First, we study Hidden Markov Models (HMMs) [4].The main issue with HMMs is to select the number of hidden states. Because of the lack of knowledge on non-human species, it is hard to have this number. This rises a model selection problem, which can be addressed by information selection criteria such as BIC, AIC [5, 6], which select an

HMM with a number of states from pre-estimated HMMs with varying number of states. Such approaches require learning multiple HMMs. On the other hand, non-parametric derivations of HMMs constitute a well-principled alternative to address this issue. Thus we used Bayesian parametric (BNP) formulation for HMMs [7], also called the infinite HMM (iHMM) [8]. It allows to infer the number of states (segments, units) from the data. The BNP approach for HMMs relies on Hierarchical Dirichlet Process (HDP) to define a prior over the states [7]. It is known as the Hierarchical Dirichlet Process for the Hidden Markov Models (HDP-HMM) [7]. The HDP-HMM parameters can be estimated by MCMC sampling techniques such as Gibbs sampling. The standard HDP-HMM Gibbs sampling has the limitation of an inadequate modeling of the temporal persistence of states [9]. This problem has been addressed by Fox et al. [9] by relying on a sticky extension which allows a more robust learning. Hence, we have a model to separate non-human signals into states that represent different activities (song units) and exploring the inference of complex data such as bioacoustic data in an environmental case (multispecies/multisources) this problem is not yet resolved.

In this paper, we investigate the BNP formulation of HMM, that is the HDP-HMM, into two challenges involving real bioacoustic data. First, a challenging problem of humpback whale song decomposition is investigated. The objective is the unsupervised structuration of whale bioacoustic data. Humpback whale songs are long cyclical sequences produced by males during the reproduction season which follows their migration from high-latitude to low-latitude waters. Singers from the same geographical region share parts of the same song. This leads to the idea of dialect [10]. Different hypotheses of these songs were emitted [11–14]. Next, we investigate a challenging problem of bird song unit structuration. Catchpole and Slater [15], Kroodsma and Miller [16] show how birds sing and why birds have such elaborate songs. However, analysing bird song units is difficult due to the transientness of typical bird chirps, the large behavioural intra-class variability, the small amount of examples per class, the presence of wildlife noise, and so forth. As shown later in the obtained segmentation results, such automatic approaches allow large-scale analysis of environmental bioacoustics recordings

## 7.1.1  Related Work

Discovering the call units (which can be considered as a kind of non-human alphabet) of such complex signals can be seen as a problem of unsupervised call units classification as [1, 17].

Picot et al. [18] also tried to analyse bioacoustic songs using a clustering approach. They implemented a segmentation algorithm based on Payne's principle to extract sound units from a bioacoustic song. Contrary to [17], in which the number of states (call units in this case) has been fixed by Davies Bouldin criteria, or [18] where a K-means algorithm is used, our approach is based on a probabilist approach on the MFCC[1]; it is non-parametric that is well-suited to the problem of automatically inferring the number of the states corresponding to the data. In the next section we describe the real-world bioacoustic challenges we used and explain our approach.

## 7.2 Data and Methods

The data used represent the difficulties of bioacoustic problems, especially when the only information linked to the signal is the species name. Thus, we have to determine a sequence without ground truth.

### 7.2.1 Humpback Whale Data

Humpback whale song data consist of a recording (about 8.6 minutes) produced at few meters from the whale in La Reunion—Indian Ocean [19],[2] at a frequency sample of 44.1 kHz, 32 bits, one channel.

We extract MFCC features from the signal, with pre-emphasis: 0.95, hamming window, FFT on 1024 points (nearly 23 ms), frameshift 10 ms, 24 Mel channels, 12 MFCC coefficients plus energy and their delta and acceleration, for a total of 39 dimensions as detailed in the NIPS 2013 challenge [19] where the signal and the features are available. The retained data for our experiment are the 51,336 first observations.

### 7.2.2 Multi-Species Bird Data

Bird species song data from Fernand Deroussen Jerome Sueur of Musee National d'Histoire Naturelle [20], consists of a training and a testing set (not used here because it contains multiple species singing simultaneously). Theses sets were designed for the ICML4B challenge.[3]

The recordings have a frequency sample of 44.1 kHz, 16 bits, one channel. The training set is composed of 35 recordings, 30 s each taken from 1 microphone. Each record contains 1 bird species in the foreground for a total of 35 different birds species.

---

[1]The MFCC are features that represent and compress short-term power spectrum of a sound. It follows the Mel scale.

[2]http://sabiod.univ-tln.fr/nips4b/challenge2.html.

[3]http://sabiod.univ-tln.fr/icml2013/BIRD_SAMPLES/.

The feature extraction for this application is applied as follows. First, a high pass filter is processed to reduce the noise (set at 1.000 kHz to avoid noises). Then, we extract the MFCC features with windows of 0.06 s and shift of 0.03 s, we keep 13 coefficients, with energy as first parameter, to be compact and sufficient accurate, considering only the vocal track information and removing the source information. Also, we focus on frequencies below 8.000 kHz, because of the alterations into the spectrum. We obtain 34,965 observations with 13 dimensions each for train set, that is used to learn our model.

### 7.2.3  Method: Unsupervised Learning for Signal Representation

To solve bioacoustic problems and finding the number of call units we propose to use the HDP-HMM model to model complex bioacoustic data. Our approach automatically discovers and infers the number of states from the non-human song data.

In this paper we present two applications on bioacoustic data. We study the song unit structuration, for the humpback whale and for the multi-species birds signal.

In the next section we give a brief description of the Hidden Markov Model and it's Bayesian non-parametric used in our bioacoustic signal representation applications.

## 7.3  Bayesian Non-parametric Alternative for Hidden Markov Model

The finite Hidden Markov Model (HMM) is very popular due to its stability to model sequential data (e.g. acoustic data). It assumes that the observed sequence $\mathbf{X} = (\mathbf{x}_1, \ldots, \mathbf{x}_T)$ is governed by a hidden state sequence $\mathbf{z} = (z_1, \ldots, z_T)$, where $\mathbf{x}_t \in \mathbb{R}^d$ is the multidimensional observation at time $t$ and $z_t$ represents the hidden state of $\mathbf{x}_t$ values in a finite set $\{1, \ldots, K\}$, $K$ being the number of states, that is unknown. The generative process of the HMM can be described in general by the following steps. First, $z_1$ follows the initial distribution $\pi_1$. Then, given the previous state $(z_{t-1})$, the current state $z_t$ follows the transition distribution. Finally, given the state $z_t$, the observation $\mathbf{x}_t$ follows the emission distribution $F(\boldsymbol{\theta}_{z_t})$ of that state. The HMM parameters, that are the initial state transition ($\pi_1$), the transition matrix ($\boldsymbol{\pi}$), and the emission parameters ($\boldsymbol{\theta}$) are in general estimated in a maximum likelihood estimation (MLE) framework by using the Expectation-Maximization (EM) algorithm, also known as the Bauch-Welch algorithm [21] in the context of HMMs.

Therefore, for the finite HMM, the number of states $K$ is required to be known a priori. This model selection issue can be addressed in a two-stage scheme by using model selection criteria such as the Bayesian Information Criterion (BIC) [5], the Akaike Information Criterion (AIC) [6], the Integrated Classification Likelihood criterion (ICL) [22], etc to select a model from pre-estimated HMMs with varying number of states. Such approaches are limited to learn $N$ HMMs, $N$ being sufficiently high to have an equivalent of a non parametric approach. In the light of this, a non parametric approach is more efficient because it theoretically tends to an infinite number of states. Thus, we use a Bayesian non-parametric (BNP) version of the HMM, that is able to infer the number of hidden states from the data. It is more flexible than learning multiple HMMs, because in bio-acoustic problems the model have to characterize multiple species/individuals, thus it possibly tends to a large number of hidden states.

The BNP approach for the HMM, that is the infinite HMM (iHMM), is based on a Dirichlet Process (DP) [23]. However, as the transitions of states have independent priors, there is no coupling across transitions between different states [8], therefore the DP is not sufficient to extend the HMM to an infinite model. The Hierarchical Dirichlet Process (HDP) prior distribution on the transition matrices over countability infinite state space, derived by Teh et al. [7], extends the HMM to the infinite state space model and is briefly described in the next subsection.

### 7.3.1 Hierarchical Dirichlet Process (HDP)

Suppose the data divided into $J$ groups, each produced by a related, yet distinct process. The HDP extends the DP by an hierarchical Bayesian approach such that a global Dirichlet Process prior $\mathrm{DP}(\alpha_0, G_0)$ is drawn from a global prior $G_j$, where $G_0$ is itself a Dirichlet Process distribution with two parameters, a base distribution $H$ and a concentration parameter $\gamma$. The generative process of the data with the HDP can be summarized as follows. Suppose data $\mathbf{X}$, with $i = 1, \ldots, T$ observations grouped into $j = 1, \ldots, J$ groups. The observations of the group $j$ are given by $\mathbf{X}_j = (\mathbf{x}_{j1}, \mathbf{x}_{j2}, \ldots)$, all observations of group $j$ being exchangeable. Assume each observation is drawn from a mixture model, thus each observations $\mathbf{x}_{ji}$ is associated with a mixture component, with parameter $\theta_{ji}$. Note that from the DP property, we observe equal values in the components $\theta_{ji}$. Now, giving the model parameter $\theta_{ji}$, the data $\mathbf{x}_{ji}$ is drawn from the distribution $F(\theta_{ji})$. Assuming a prior distribution $G_j$ over the model parameters associated for group $j$, $\boldsymbol{\theta}_j = (\theta_{j1}, \theta_{j2}, \ldots)$, we can define the generative process in Eq. (7.1).

$$
\begin{aligned}
G_0|\gamma, H &\sim \mathrm{DP}(\gamma, H), \\
G_j|\alpha_0, G_0 &\sim \mathrm{DP}(\alpha_0, G_0), \ \forall j \in 1, \ldots, J, \\
\theta_{ji}|G_j &\sim G_j, \ \forall j \in 1, \ldots, J \text{ and } \forall i \in 1, \ldots, T, \\
\mathbf{x}_{ji}|\theta_{ji} &\sim F(\mathbf{x}_{ji}|\theta_{ji}), \forall j \in 1, \ldots, J \text{ and } \forall i \in 1, \ldots, T.
\end{aligned}
\tag{7.1}
$$

The Chinese Restaurant Process (CRP) [24] is a representation of the Dirichlet Process that results from a metaphor related to the existence of a restaurant with possible infinite tables (clusters) where customers (observations) are sitting in it. An alternative of such a representation for the Hierarchical Dirichlet Process can be described by the Chinese Restaurant Franchise (CRF) process by extending the CRP to multiple restaurants that share a set of dishes.

The idea of CRF is that it gives a representation for the HDP by extending a set of (J) restaurants, rather than a single restaurant. Suppose a patron of chinese restaurant creates many restaurants, strongly linked to each other, by a franchise wide menu, having dishes common to all restaurants. As a result, restaurants are created (groups) with a possibility to extend each restaurant with an infinite number of tables (states) at witch the customers (observations) sit. Each customer goes to his specified restaurant $j$, where each table of this restaurant has a dish between the customers that sit at that specific table. However, multiple tables of different existing restaurants can serve the same dish.

### 7.3.2  The Hierarchical Dirichlet Process for the Hidden Markov Model (HDP-HMM)

The HDP-HMM uses a HDP prior distribution providing a potential countability infinite number of hidden states and tackles the challenging problem of model selection for the HMM. This model is a Bayesian non-parametric extension for the HMM also presented as the infinite HMM. To derive the HDP-HMM model we suppose a doubly-infinite transition matrix, where each row corresponds to a CRP. Thus, in a HDP formalism, the groups correspond to states, with CRP distribution on next states. CRF links these states distributions.

We assume for simplicity a distinguished initial state $z_0$. Let $G_j$ describes both, the transition matrix $\pi_k$ and the emission parameters $\theta_k$, the infinite HMM can be described by the following generative process:

$$\beta|\gamma \sim \text{GEM}(\gamma),$$

$$\pi_k|\alpha, \beta \sim \text{DP}(\alpha, \beta),$$

$$z_t|z_{t-1} \sim \text{Mult}(\pi_{z_{t-1}}), \tag{7.2}$$

$$\theta_k|H \sim H,$$

$$\mathbf{x}_t|z_t, \{\theta_k\}_{k=1}^\infty \sim F(\theta_{z_t}).$$

where,

$\beta$ is a hyperparameter for the DP [2] distributed according to the stick-breaking construction noted GEM(.);

$z_t$ is the indicator variable of the HDP-HMM that follows a multinomial distribution Mult(.);

the emission parameters $\boldsymbol{\theta}_k$, are drawn independently, according to a conjugate prior distribution $H$;

$F(\boldsymbol{\theta}_{z_t})$ is a data likelihood density with the unique parameter space of $\boldsymbol{\theta}_{z_t}$ equal to $\boldsymbol{\theta}_k$.

Suppose the observed data likelihood is a Gaussian density $\mathcal{N}(\mathbf{x}_t; \boldsymbol{\theta}_k)$ where the emission parameters $\boldsymbol{\theta}_k = \{\boldsymbol{\mu}_k, \boldsymbol{\Sigma}_k\}$ are respectively the mean vector $\boldsymbol{\mu}_k$ and the covariance matrix $\boldsymbol{\Sigma}_k$. According to [25], the prior over the mean vector and the covariance matrix is a conjugate Normal-Inverse-Wishart distribution, denoted as $\mathcal{N}\mathcal{I}\mathcal{W}(\mu_0, \kappa_0, \nu_0, \Lambda_0)$, with the hyper-parameters describing the shapes and the position for each mixture components: $\mu_0$ is the mean of Gaussian should be, $\kappa_0$ the number of pseudo-observations supposed to be attributed, and $\nu_0, \Lambda_0$ being similarly for the covariance matrix.

In the generative process given in Eq. (7.2), $\boldsymbol{\pi}$ is interpreted as a double-infinite transition matrix with each row taking a CRP. Thus, in the HDP formulation the group-specific distribution, $\boldsymbol{\pi}_k$ corresponds to the state-specific transition where the CRF defines distributions over the next state. In turn, [9] showed that HDP-HMM inadequately models the temporal persistence of states, creating redundant and rapidly switching states and proposed an additional hyperparameter $\kappa$ that increase the self-transition probabilities. This is named as sticky HDP-HMM. The distribution on the transition matrix of Eq. (7.2) for the sticky HDP-HMM is given as follows:

$$\boldsymbol{\pi}_k|\alpha, \boldsymbol{\beta} \sim \mathrm{DP}\left(\alpha + \kappa, \frac{\alpha\boldsymbol{\beta} + \kappa\delta_k}{\alpha + \kappa}\right), \tag{7.3}$$

where a small positive $\kappa > 0$ is added to the $k^{\text{th}}$ component of $\alpha\boldsymbol{\beta}$, thus of self-transition probability is increased by $\kappa$. Note that setting $\kappa$ to 0, the original HDP-HMM is recovered. Under such assumption for the transition matrix, [9] proposes an extension of the CRF to the Chinese Restaurant Franchise with Loyal Customers. A graphical representation of (sticky) HDP-HMM is given in Fig. 7.2.

**Fig. 7.2** Graphical representation of sticky Hierarchical Dirichlet Process for Hidden Markov Model (HDP-HMM)

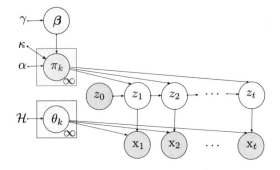

The inference of the infinite HMM (the (sticky) HDP-HMM) with the Block Gibbs sampler algorithm is given in Algorithm 3 of Supplementary Material in [9] paper. The basic idea of this sampler is to estimate the posterior distributions over all the parameters from the generative process of (sticky) HDP-HMM given in Eq. (7.2). Here, the CRF with loyal customers, hyperparameter $\kappa$ of the transition matrix can be sampled in order to increase the self-transition probability.

Hence, the HDP-HMM model resolves the problem of advanced signal decomposition using acoustic features with respect to time. It allows identifying song units (states), behaviour and enhancing populations studies. From the other point, modelling data with the HDP-HMM offers a great alternative of the standard HMM to tackle the challenging problem of selecting the number of states, identifying the unknown number of hidden units from the used features (here: MFCC). The experimental results show the interest of such an approach.

## 7.4  Experiments

In this section we present two applications on bioacoustic data. We study the song unit structuration, for the humpback whale signal and for multi-species birds signals.

### 7.4.1  Humpback Whale Sound Segmentation

The learning of the humpback whale song, applied via the HDP-HMM, is done with the Blocked Gibbs sampling. A number of iterations was fixed to $N_s = 30,000$ and a truncation level, that corresponds to the maximum number of possible states in the model (being sufficient big to approximate it to an infinite model), is fixed to $L_k = 30$. The number of states estimated by the HDP-HMM Gibbs sampling is six.

Figure 7.3 shows the state sequences partition, for all 8.6 min of humpback whale song data, obtained by the HDP-HMM Gibbs sampling. For more detailed information, the result of the whole humpback whale signal segmentation is separated by several parts of 15 s. All the spectrograms of the humpback whale song and the obtained segmentation are made available in the demo: http://sabiod.univ-tln.fr/workspace/MTAP/whale.zip. This demo highlights the interest of using a BNP formulation of HMMs for unsupervised segmentation of whale signals. Three examples of the humpback whale song, with 15 s duration each, are presented and discussed in this paper (see Fig. 7.5).

Figure 7.5 represents the spectrogram and the corresponding state sequence partition obtained by the HDP-HMM Gibbs inference algorithm. They respectively represent examples of the beginning, the middle and the end of the whole signal.

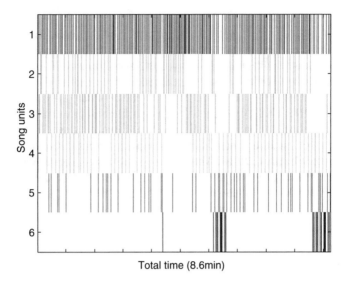

**Fig. 7.3** State sequence for 8.6 min of humpback whale song obtained by the Blocked Gibbs sampling inference approach for HDP-HMM

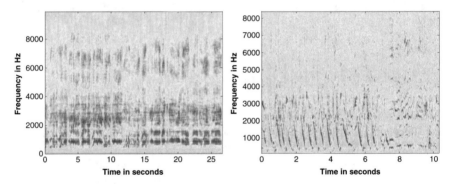

**Fig. 7.4** Spectrograms of the 6th whale song unit (left) and 2nd song unit (right)

All the obtained state sequence partitions fit the spectral patterns. We note that the estimated state 1 fits the sea noise, state 5 also fits sea noise, but it is right before units associated to whale songs. The presence of this unit can be due to an insufficient number of Gibbs samples. For a longer learning the fifth state could be merged with the first state. State 2 fits the up and down sweeps. State 3 fits low and high fundamental harmonic sounds, state 4 fits for numerous harmonics sound and state 6 fits very noisy and broad sounds. Figure 7.4 shows two spectrograms extracted from the 6th song unit (left) and from the 2nd song unit (right) of the whole humpback whale signal. We can see that the units fit specific patterns on the whole signal.

Pr. Gianni Pavan (Pavia University, Italy), undersea NATO bioacoustic expert analysed the results on these humpback whale song segmentations we produced

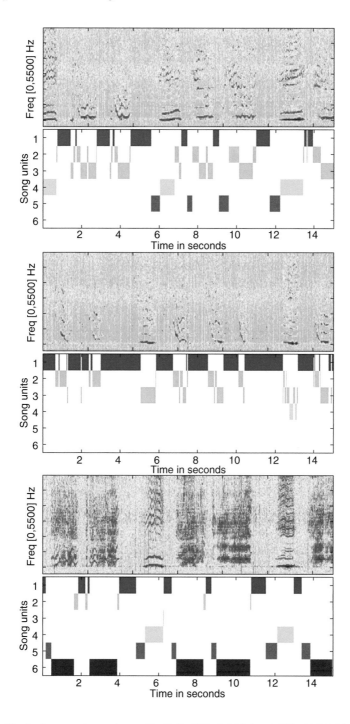

**Fig. 7.5** Obtained song units starting at 60 s (left), 255 s (middle) and 495 s (right). The spectrogram of the whale song (top), and the obtained state sequence (bottom) by the Blocked Gibbs sampler inference approach for the HDP-HMM. The silence (unit 1 and 5) looks well separated from the whale signal. Whale up and down sweeps (unit 2), harmonics (unit 3 and 4) and broad sounds (unit 6) are also present

in this paper. He validated the computed representation, as the usual optimal segmentation an expert produces. This highlight the interest of learning BNP model on a single species to produce expert representation. In the next section we validate the approach on several bird species.

## 7.4.2   Birds Sound Segmentation

In this section we describe the obtained bird song unit segmentation. We segment the bird signals into song units by learning the HDP-HMM model on the training set (containing 35 different species). The main goal is to see if a such approach can model multiple species. Note that in this set, we assume there is no multiple species singing at the same time.

For this application, we considered 14,5000 Gibbs iterations and a truncation level of 200 for the maximum number of states. We suppose them to be sufficiently big for this data problem. Moreover, we use one mixture component per state, that appeared to give satisfactory results and we use a sticky HDP-HMM with the hyper-parameter $\kappa$ set to 0.1.

We discovered 76 song units with this method. For more detailed information over the signal, we separated the whole train set into parts of 15 s each. All the spectrograms and the associated segmentation obtained are made available in the demo: http://sabiod.univ-tln.fr/workspace/MTAP/bird.zip.

### 7.4.2.1   Evaluation of the Bird Result

To evaluate the bird results, we used a ground truth produced by an expert ornithologist. He segmented each recording of the dataset according to the different patterns on the signal. Then we compare this ground truth with the segments produced by the model using the Normalized Mutual Information NMI [26] which calculates shared information between two clustering sets. We computed the NMI score for each species, as reported in Table 7.1. The highest score is 0.680 (*Corvus Corone*) and the lowest score is 0.003 (*Garrulus Glandarius*). Thus, for some species, the model has difficulties to segment the data. Sometimes, it uses less states than the expert to segment the data: for the *Oriolus Oriolus* (*Golden Oriole*), the model identifies 12 song units versus 50 identified by the expert. Nevertheless, the model also uses more states than the expert to segment the data: for the *Fringilla Coelebs* (chaffinch), the model identifies 15 song units versus 3 identified by the expert. In other cases, the model can't differentiate 2 distinct vocalizes if they have close frequencies (*Phylloscopus Collybita* and *Columba Palumbus*), background and foreground species (*Streptopelia Decaocto*). This can be due to the feature used (wrong time scale), or to an insufficient number of iterations of the Gibbs sampling. For most of species, the model and the ground truth have similar patterns observable on Figs. 7.6, 7.8 and 7.7, but not in the sample Figs. 7.10 and 7.9.

**Table 7.1** NMI score for the obtained segmentation using HDP-HMM

| Species | NMI score |
|---|---|
| *Corvus corone* | 0.680 |
| *Picus viridis* | 0.602 |
| *Fringilla coelebs* | 0.565 |
| *Emberiza citrinella* | 0.534 |
| *Parus palustris* | 0.521 |
| *Luscinia megarhynchos* | 0.497 |
| *Dendrocopos major* | 0.481 |
| *Prunella modularis* | 0.476 |
| *Sturnus vulgaris* | 0.467 |
| *Pavo cristatus* | 0.437 |
| *Certhia brachydactyla* | 0.417 |
| *Turdus viscivorus* | 0.417 |
| *Parus caeruleus* | 0.413 |
| *Troglodytes troglodytes* | 0.407 |
| *Sylvia atricapilla* | 0.405 |
| *Turdus philomelos* | 0.398 |
| *Turdus merula* | 0.395 |
| *Erithacus rubecula* | 0.394 |
| *Carduelis chloris* | 0.385 |
| *Columba palumbus* | 0.352 |
| *Branta canadensis* | 0.339 |
| *Anthus trivialis* | 0.332 |
| *Sitta europaea* | 0.332 |
| *Oriolus oriolus* | 0.316 |
| *Streptopelia decaocto* | 0.306 |
| *Phoenicurus phoenicurus* | 0.291 |
| *Phasianus colchicus* | 0.272 |
| *Parus major* | 0.270 |
| *Phylloscopus collybita* | 0.267 |
| *Cuculus canorus* | 0.205 |
| *Aegithalos caudatus* | 0.202 |
| *Strix aluco* | 0.200 |
| *Alauda arvensis* | 0.169 |
| *Motacilla alba* | 0.105 |
| *Garrulus glandarius* | 0.003 |
| Mean | 0.367 |

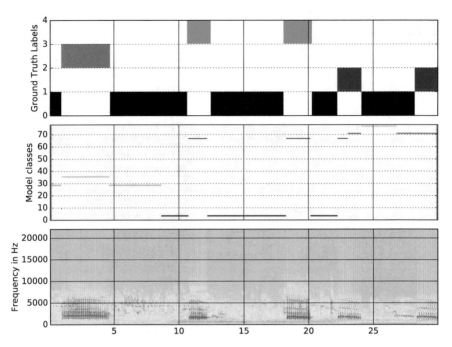

**Fig. 7.6** Picus viridis with a high NMI score of 0.602. Top: the labelled ground truth over 30 s where label 0 is always the silence label and the other labels are specific to each species. Medium: our model with the 76 classes. Bottom: spectrogram

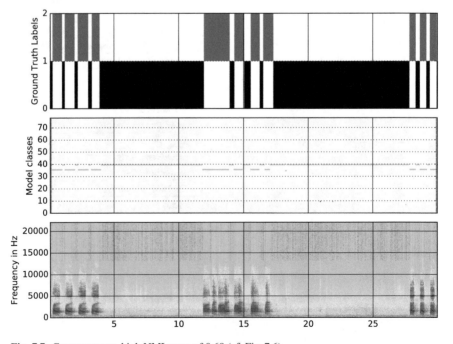

**Fig. 7.7** Corvus corone, high NMI score of 0.68 (cf. Fig. 7.6)

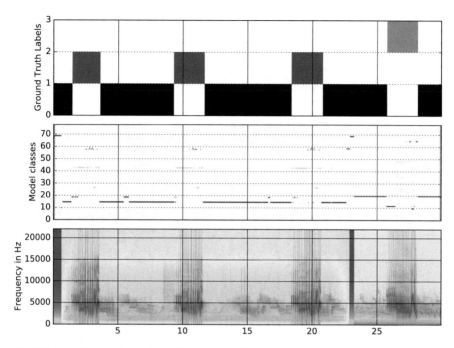

**Fig. 7.8** Fringilla coelebs, medium NMI score 0.565 (cf. Fig. 7.6)

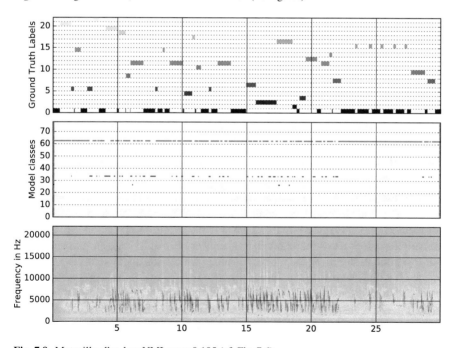

**Fig. 7.9** Motacilla alba, low NMI score 0.105 (cf. Fig. 7.6)

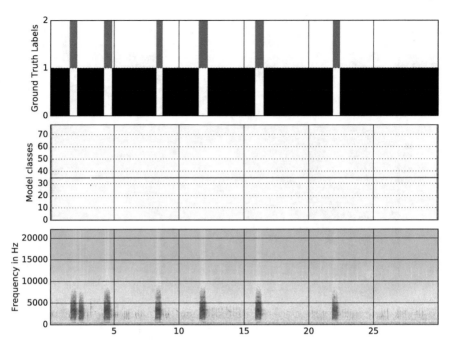

**Fig. 7.10** Garrulus glandarius, low NMI score 0.003 (cf. Fig. 7.6)

To improve the model, we can investigate better feature representation for species with different acoustic characteristics. We can also improve noise reduction which could be useful for background activities. Also, it can be dur to the fact we use one annotator. Nevertheless, the application highlights the interest of using BNP formulation of HMMs for unsupervised segmentation of bird signals.

## 7.5  Conclusions

We proposed BNP HMM formulation to a representation of real world bioacoustic scenes. The evaluations on two challenges, available online, show the efficiency of the method, which forms a possible answer to the questions opened in [3]. The BNP formulation gives an estimate number of cluster needed to segment the signal and our experiments highlight the interest of such formulation on bioacoustic problems. We score with NMI the segmentation obtained for birds with the segmentation from an expert, showing promising results.One of the main topic in ecological acoustics is the development of unsupervised methods for automatic detection of vocalized species, which would help specialists in ecological works during their monitoring activities.Future work will consist in the MCMC sampling dealing with larger data problems, like variational inference [27] or stochastic variational inference used for HMMs [28], joint to feature learning to automatically adapt time frequency scales to each species.

**Acknowledgements** We would like to thanks Provence-Alpes-Côte d'Azur region and NortekMed for their financial support for Vincent ROGER. We also thank GDR CNRS MADICS http://sabiod.org/EADM for its support. We thank G. Pavan for its expertise, J. Sueur, F. Deroussen, F. Jiguet for the coorganisation of the challenges and M. Roch for her collaboration.

# References

1. Bartcus, M., Chamroukhi, F., & Glotin, H. (2015, July). Hierarchical Dirichlet Process Hidden Markov Model for Unsupervised Bioacoustic Analysis. In Neural Networks (IJCNN), 2015 International Joint Conference on pp. 1–7. IEEE.
2. Sethuraman, J. (1994). A constructive definition of Dirichlet priors. Statistica sinica, 639–650.
3. Kershenbaum, A., Blumstein, D.T., Roch, M.A., Akçay, Ç., Backus, G., Bee, M.A., Bohn, K., Cao, Y., Carter, G., Cäsar, C. and Coen, M. (2016). Acoustic sequences in non-human animals: a tutorial review and prospectus. Biological Reviews, 91(1), pp.13–52.
4. Rabiner, L. and Juang, B. (1986). An introduction to hidden Markov models. ieee assp magazine, 3(1), pp.4–16.
5. Schwarz, G. (1978). Estimating the dimension of a model. The annals of statistics, 6(2), pp.461–464.
6. Akaike, H. (1974). A new look at the statistical model identification. IEEE transactions on automatic control, 19(6), 716–723.
7. Teh, Yee Whye and Jordan, Michael I. and Beal, Matthew J. and Blei, David M. (2006). Hierarchical Dirichlet Processes. Journal of the American Statistical Association, 476(101), pp.1566–1581.
8. Beal, M. J., Ghahramani, Z., & Rasmussen, C. E. (2002). The infinite hidden Markov model. In Advances in neural information processing systems pp. 577–584.
9. Fox, E. B., Sudderth, E. B., Jordan, M. I., & Willsky, A. S. (2008, July). An HDP-HMM for systems with state persistence. In Proceedings of the 25th international conference on Machine learning pp. 312–319. ACM.
10. Helweg, D.A., Cat, D.H., Jenkins, P.F., Garrigue, C. and McCauley, R.D. (1998). Geograpmc Variation in South Pacific Humpback Whale Songs. Behaviour, 135(1), pp.1–27.
11. Medrano, L., Salinas, M., Salas, I., Guevara, P.L.D., Aguayo, A., Jacobsen, J. and Baker, C.S. (1994). Sex identification of humpback whales, *Megaptera novaeangliae*, on the wintering grounds of the Mexican Pacific Ocean. Canadian journal of zoology, 72(10), pp.1771–1774.
12. Frankel, A.S., Clark, C.W., Herman, L. and Gabriele, C.M. (1995). Spatial distribution, habitat utilization, and social interactions of humpback whales, *Megaptera novaeangliae*, off Hawai'i, determined using acoustic and visual techniques. Canadian Journal of Zoology, 73(6), pp.1134–1146.
13. Baker, C.S. and Herman, L.M. (1984). Aggressive behavior between humpback whales (*Megaptera novaeangliae*) wintering in Hawaiian waters. Canadian journal of zoology, 62(10), pp.1922–1937.
14. Garland, E.C., Goldizen, A.W., Rekdahl, M.L., Constantine, R., Garrigue, C., Hauser, N.D., Poole, M.M., Robbins, J. and Noad, M.J. (2011). Dynamic horizontal cultural transmission of humpback whale song at the ocean basin scale. Current Biology, 21(8), pp.687–691.
15. Catchpole, C.K. and Slater, P.J., 86. B. (1995). Birdsong: Biological Themes and Variations. Cambridge University PressCatchpole.

16. Kroodsma, D. E., & Miller, E. H. (Eds.). (1996). Ecology and evolution of acoustic communication in birds pp. 269–281. Comstock Pub.
17. Pace, F., Benard, F., Glotin, H., Adam, O. and White, P. (2010). Subunit definition and analysis for humpback whale call classification. Applied Acoustics, 71(11), pp.1107–1112.
18. Picot, G., Adam, O., Bergounioux, M., Glotin, H. and Mayer, F.X. (2008, October). Automatic prosodic clustering of humpback whales song. In New Trends for Environmental Monitoring Using Passive Systems, 2008 pp. 1–6. IEEE.
19. Glotin, H., LeCun, Y., Artieres, T., Mallat, S., Tchernichovski, O., & Halkias, X. (2013). Neural information processing scaled for bioacoustics, from neurons to big data. USA (2013). http://sabiod.org/NIPS4B2013_book.pdf.
20. Deroussen F., Jiguet F. (2006). La sonotheque du Museum: Oiseaux de France. Nashvert Production, Charenton, France.
21. Baum, L.E., Petrie, T., Soules, G. and Weiss, N. (1970). A maximization technique occurring in the statistical analysis of probabilistic functions of Markov chains. The annals of mathematical statistics, 41(1), pp.164–171.
22. Biernacki, C., Celeux, G. and Govaert, G. (2000). Assessing a mixture model for clustering with the integrated completed likelihood. IEEE transactions on pattern analysis and machine intelligence, 22(7), pp.719–725.
23. Ferguson, T.S. (1973). A Bayesian analysis of some nonparametric problems. The annals of statistics, pp.209–230.
24. Pitman, J. (1995). Exchangeable and partially exchangeable random partitions. Probability theory and related fields, 102(2), pp.145–158.
25. Gelman, A., Carlin, J. B., Stern, H. S., & Rubin, D. B. (2003). Bayesian Data Analysis, (Chapman & Hall/CRC Texts in Statistical Science).
26. Strehl, A. and Ghosh, J. (2002). Cluster ensembles—a knowledge reuse framework for combining multiple partitions. Journal of machine learning research, 3(Dec), pp.583–617.
27. Jordan, M.I., Ghahramani, Z., Jaakkola, T.S. and Saul, L.K. (1999). An introduction to variational methods for graphical models. Machine learning, 37(2), pp.183–233.
28. Foti, N., Xu, J., Laird, D., & Fox, E. (2014). Stochastic variational inference for hidden Markov models. In Advances in neural information processing systems, pp.3599–3607.

# Chapter 8
# Plant Identification: Experts vs. Machines in the Era of Deep Learning

## Deep Learning Techniques Challenge Flora Experts

Pierre Bonnet, Hervé Goëau, Siang Thye Hang, Mario Lasseck, Milan Šulc, Valéry Malécot, Philippe Jauzein, Jean-Claude Melet, Christian You, and Alexis Joly

**Abstract** Automated identification of plants and animals have improved considerably in the last few years, in particular thanks to the recent advances in deep learning. The next big question is how far such automated systems are from the human expertise. Indeed, even the best experts are sometimes confused and/or disagree between each others when validating visual or audio observations of living organism. A picture or a sound actually contains only a partial information that is usually not sufficient to determine the right species with certainty. Quantifying this uncertainty and comparing it to the performance of automated systems is of high interest for both computer scientists and expert naturalists. This chapter reports an experimental study following this idea in the plant domain. In total, nine deep-learning systems implemented by three different research teams were evaluated with regard to nine expert botanists of the French flora. Therefore, we created a small set of plant observations that were identified in the field and revised by experts in

---

P. Bonnet (✉) · H. Goëau
CIRAD, UMR AMAP, Montpellier, France

AMAP, Univ Montpellier, CIRAD, CNRS, INRA, IRD, Montpellier, France
e-mail: pierre.bonnet@cirad.fr; herve.goeau@cirad.fr

S. T. Hang
Toyohashi University of Technology, Toyohashi, Japan
e-mail: hang@kde.cs.tut.ac.jp

M. Lasseck
Museum Fuer Naturkunde Berlin, Leibniz Institute for Evolution and Biodiversity Science, Berlin, Germany
e-mail: Mario.Lasseck@mfn-berlin.de

M. Šulc
Czech Technical University in Prague, Prague, Czech Republic
e-mail: sulcmila@cmp.felk.cvut.cz

© Springer International Publishing AG, part of Springer Nature 2018
A. Joly et al. (eds.), *Multimedia Tools and Applications for Environmental & Biodiversity Informatics*, Multimedia Systems and Applications,
https://doi.org/10.1007/978-3-319-76445-0_8

order to have a near-perfect golden standard. The main outcome of this work is that the performance of state-of-the-art deep learning models is now close to the most advanced human expertise. This shows that automated plant identification systems are now mature enough for several routine tasks, and can offer very promising tools for autonomous ecological surveillance systems.

## 8.1   Introduction

Automated species identification was presented 15 years ago as a challenging but very promising solution for the development of new research activities in Taxonomy, Biology or Ecology [17]. With the development of an increasing number of web and mobile applications based on visual data analysis, the civil society was able in the recent years to evaluate the progress in this domain, and to provide new data for the development of large-scale systems [15]. To evaluate the performance of automated plant identification technologies in a sustainable and repeatable way, a dedicated system-oriented benchmark was setup in 2011 in the context of the CLEF evaluation forum [21]. A challenge called PlantCLEF was organized in this context using datasets co-produced with actors of the civil society (such as educators, nature lovers, hikers). Years after years, the complexity and size of this testbed was increasing and allowed dozens of research teams to evaluate the progress and limits of the machine learning systems they developed. In 2017, the PlantCLEF challenge was organized on a dataset covering 10,000 plant species. This was the first evaluation at this scale in the world, and results were promising and impressive with accuracies reaching 90% of correct identification for the best system. This amazingly high performance raises the question of how far automated systems are from the human expertise and of whether there is a upper bound that can not be

V. Malécot
IRHS, Agrocampus-Ouest, Rennes, France

INRA, Université d'Angers, Angers, France
e-mail: valery.malecot@agrocampus-ouest.fr

P. Jauzein
AgroParisTech UFR Ecologie Adaptations Interactions, Paris, France
e-mail: p.jauzein@free.fr

J.-C. Melet
Independent Botanist
e-mail: jcd.melet@wanadoo.fr

C. You
Société Botanique Centre Ouest, Nercillac, France
e-mail: you.christian@neuf.fr

A. Joly
Inria ZENITH Team, Montpellier, France
e-mail: alexis.joly@inria.fr

exceeded. A picture (or a set of pictures) actually contains only a partial information about the observed plant and it is often not sufficient to determine the right species with certainty. For instance, a decisive organ such as the flower or the fruit, might be not visible at the time the plant was observed. Or some of the discriminant patterns might be very hard or unlikely to be observed in a picture such as the presence of hairs or latex, or the morphology of the underground parts. As a consequence, even the best experts can be confused and/or disagree between each others when attempting to identify a plant from a set of pictures. Estimating this intrinsic data uncertainty according to human experts and comparing it to the performance of the best automated systems is of high interest for both computer scientists and expert naturalists.

A first step in that direction had been taken in 2014 through a first *Man vs. Machine experiment* conducted by some of the authors of this chapter [1]. At that time, it was concluded that machines were still far from performing as well as expert botanists. The best methods were only able to outperform the participants that declared themselves as amateurs of botany. Computer vision has made great progress since that time, in particular thanks to the advances in deep learning. Thus, this chapter presents an upgraded *Human vs. Machine* experiment in the continuity of the previous study but using state-of-the-art deep learning systems. For a fair comparison, we also extended the evaluation dataset to more challenging species and we involved expert botanists with a much higher expertise on the targeted flora. In total, nine deep-learning systems implemented by three different research teams were evaluated with regard to nine expert botanists among the most renowned in Europe. The rest of this chapter is organized as follows. In Sect. 8.2, we first return to the process of identifying a plant by an expert in order to fully understand its mechanisms and issues. Then, in Sect. 8.3, we give an overview of the state-of-the-art in automated plant identification by synthesizing the results of the international evaluation campaign LifeCLEF 2017 co-organized by some of the authors of this chapter. Finally, in Sect. 8.4.2, we report the results and analysis of our new *experts vs. machines* experiment.

## 8.2 Understanding the Plant Identification Process by Botanists

For a botanist, identifying a plant means associating a scientific name to an individual plant. More precisely, that means assigning that individual plant to a group, called a taxon. Such taxon had a name selected according to a set of rules. The delimitation of taxa and the scientific names applying to them are the result of a process called taxonomy (or systematics). This process is in the hands of a relatively low number of scientists. During that process, hundreds of herbarium sheets (i.e. dry plants collected during the past centuries and mounted on a large piece of paper together with annotations such as date, place, collector name...) and usually a lower number of living plants are compared. Such comparison may be based on macromorphological, micromorphological or molecular data, manually or computationally analyzed. This comparison allows delimiting groups on the basis of certain features. This is a step where the taxonomist should tell apart variability in the morphology of the various parts of the individuals assigned to a peculiar taxa and features shared by all the specimens assigned to that taxa. The obtained groups are hierarchically organized, in a classification. The most common rank in such classifications is the species, but other ranks are used such as genus, family. Thus, identifying a plant is commonly treated as giving the scientific name at the specific rank. To do this, botanists relies on various methods involving memory and observation. As the result of a more or less long learning, botanist may have an implicit knowledge of the appearance and the variability of a species. Botanists may also rely on diagnostic characters, i.e. features (morphological) that tell apart individual of a peculiar species from any other species in an area. For example, any fan-like leaf, with a median sinus, collected on a tree, may be assigned to *Ginkgo biloba* (among living plants). Diagnostic characters may also correspond to some higher ranked taxa, for example, umbel-like inflorescences of *Apiaceae*. Additionally, botanists may also use identification keys. Such tools consist in a set of alternatives, usually a pair of morphological characters (for example "leaf less than 10 cm long" versus "leaf equal or more than 10 cm long"). At each set of alternatives the botanist should select the morphological character best applying to his sample. This drives him toward another set of alternative or to the name of his material. Production of identification keys is a complex process, and, when allowing the identification of the plants of an area or a large taxonomic group (such as a family or a genus), are assembled in books called Floras or Monographs. Such published paper material is generally used by professional botanists, students, land managers or nature observers in general.

In the field, expert botanists may apply more or less simultaneously the three above-listed methods, i.e. implicit knowledge, diagnostic characters and keys. Further elements may also be involved in the identification process.

1. According to the period of the year, the location, the altitude, and the local environment (such as the level of sun exposure, the distance to a river stream or a disturbed area, the soil quality, etc.), the botanist will have in mind a selection of potential plant species that occur in the prospected area. The size and the quality of this potential species list will be directly related to his/her expertise and experience on this flora.

2. When a botanist sees one or several specimens of the same species to be identified, he/she will first selects the one(s) that appear(s) to be the most informative, e.g. the most healthy, the one with the higher number of reproductive organs (flowers and fruits), or vegetative parts (stems, leaves). Due to this selection, he/she will access to the plant that will have the most higher volume of information, and that gives the best chances to lead to a correct identification.

3. Whether or not he/she uses a key, he/she may look attentively at several parts of the plants. The habit, i.e the shape of the whole plant, will then usually be the first morphological attribute analyzed by the botanist, simply because it can be seen the farthest.The flowers and the fruits, if present, are also very regularly

observed, as they are the most informative parts of the plant. Several attributes will be analyzed such as their position and insertion on the plants, their number, density, size, shape, structure, etc. Unfortunately, most plants are in flowers and fruits only a small fraction of the year (from few days to few weeks). In such situation, it is often necessary to analyze dry or dead flowers or fruits, if present. Regarding vegetative parts, most of the time, leaves are the first part to be analyzed. The botanist may examine their position and distribution along the stem, their shape, color, vein network, pubescence, etc. He/she will also try to observe uncommon particularities on the plant such as the presence of spines, of swollen parts, if some latex is flowing from the stem, or if the plant has a specific smell, etc.

4. The number of observed attributes is very variable from one plant to another. It depends on its growing stage, on the number of its morphological similar relatives for the considered flora, and of the expertise of the botanist. For example, in Europe, if a botanist identifies a specimen as belonging to the *Moraceae* family (based on the analysis of the leaf, fruit, and latex), he already knows that the number of potential species if very small. He/she doesn't have to look to many more characters for its species identification. On the other hand, if he/she identifies a specimen as a representative of the *Poaceae* family (based on the analysis of the fruits), he will have to look to many different characters as this family is one of the most rich in temperate regions (with hundreds or thousands of species).

5. If using a key, the botanist will look more precisely on the features considered at each set of alternatives, following the order used in the key (thus going from one part to another and back to the first for example). If he news and recognize on his sample diagnostic features applying to a group of species (genus, family for example), he may goes directly to the part of the key dealing with that group. If he had implicit knowledge of the plant at hand, he may use the key in a reverse way. In such situation he will goes to the set of alternatives that ends with the species' name he had in mind, and look at the characters that are used in the few previous set of alternatives. Whatever the botanist select himself the characters to look at or follows the order imposed by the key, for the same character (for example number of petals) the botanist will look at several relevant parts of the plant (in the example, several flowers), or even to several individuals, in order to prevent him looking at an anomaly.

6. During the whole identification process, botanists often use micro-lens. This allows them observing very small plant parts such as the inner parts of the flowers, or the hair shape on the leaf surface.

7. They may bring back to their offices specimens who are not easily identifiable in the field either because of lack of some characters or because of the size of such characters. They may also bring back specimen which are the most interesting for their research subject for further comparison with previously identified material.

The identification process in the field allows to better understand the assets and limits of an image-based identification. A picture (or a set of pictures) only provides a partial view of all the attributes that can be observed in the field. Indeed, the degree of informativeness of an observation is itself highly dependent on the botanical expertise of the photographer. Observations made by novices might for instance be restricted to the habit view which makes the identification impossible in some cases. Furthermore, the image-based identification process cannot be as iterative and dynamic as in the field. If the botanist realizes that an attribute is missing when following a dichotomous key, he cannot return to the observation of the plant.

## 8.3    State-of-the-Art of Automated Plant Identification

To evaluate the performance of automated plant identification technologies in a sustainable and repeatable way, a dedicated system-oriented benchmark was setup in 2011 in the context of ImageCLEF.[1] Between 2011 and 2017, about 10 research groups participated yearly to this large collaborative evaluation by benchmarking their image-based plant identification systems (see [10, 12, 13, 18–21] for more details). The last edition, in 2017, was an important milestone towards building systems working at the scale of a continental flora [10]. To overcome the scarcity of expert training data for many species, the objective was to study to what extent a huge but very noisy training set collected through the Web is competitive compared to a relatively smaller but trusted training set checked by experts. As a motivation, a previous study conducted by Krause et al. [11] concluded that training deep neural networks on noisy data was very effective for fine-grained recognition tasks. The PlantCLEF 2017 challenge completed their work in two main points: (1) it extended it to the plant domain and (2), it scaled the comparison between clean and noisy training data to 10 K of species. In the following subsections, we synthesize the methodology and main outcomes of this study. A more detailed description and a deeper analysis of the results can be found in [10].

### 8.3.1    Dataset and Evaluation Protocol

Two large training data sets both based on the same list of 10.000 plant species (living mainly in Europe and North America) were provided:

**Trusted Training Set** *EoL10K*   A trusted training set based on the online collaborative Encyclopedia Of Life (EoL).[2] The 10 K species were selected as the most populated species in EoL data after a curation pipeline (taxonomic alignment, duplicates removal, herbarium sheets removal, etc.).

---

[1]www.imageclef.org.

[2]http://eol.org/.

**Noisy Training Set *Web10K*** A noisy training set built through Web crawlers (Google and Bing image search engines) and containing 1.1 M images.

The main idea of providing both datasets was to evaluate to what extent machine learning and computer vision techniques can learn from noisy data compared to trusted data (as usually done in supervised classification). Pictures of EoL are themselves coming from several public databases (such as Wikimedia, Flickr, iNaturalist) or from some institutions or less formal websites dedicated to botany. All that pictures can be potentially revised and rated on the EoL website. On the other side, the noisy set contained more images for a lot of species, but with several types and levels of noise which are basically impossible to automatically filter: a picture can be associated to the wrong species but the correct genus or family, a picture can be a portrait of a botanist working on the species, the pictures can be associated to the correct species but be a drawing or an herbarium sheet of a dry specimen, etc.

**Mobile Search Test Set** The test data to be analyzed was a large sample of the query images submitted by the users of the mobile application Pl@ntNet (iPhone[3] and Android[4]). It contained a large number of wild plant species mostly coming from the Western Europe Flora and the North American Flora, but also species used all around the world as cultivated or ornamental plants.

## 8.3.2   *Evaluated Systems*

Eight research groups participated to the evaluation. Details of the methods and systems they used are synthesized in the overview of the task [10] and further developed in the individual working notes of the participants (CMP [2], FHDO BCSG [3], KDE TUT [4], Mario MNB [5], Sabanci Gebze [6], UM [7] and UPB HES SO [8]). Participants were allowed to run up to four systems or four different configurations of their system. In total, 29 systems were evaluated. We give hereafter more details of the techniques and methods used by the three participants who developed the best performing systems:

**Mario TSA Berlin, Germany, 4 Runs, [5]** This participant used ensembles of fine-tuned CNNs pre-trained on ImageNet [14] based on three architectures (GoogLeNet, ResNet-152 and ResNeXT) each trained with bagging techniques. Intensive data augmentation was used to train the models with random cropping, horizontal flipping, variations of saturation, lightness and rotation. Test images were also augmented and the resulting predictions averaged. *MarioTsaBerlin Run 1* results from the combination of the three architectures trained on the trusted datasets only (EOL and PlantCLEF2016). Run 2 exploited both the trusted and the noisy

---

[3]https://itunes.apple.com/fr/app/plantnet/id600547573?mt=8.

[4]https://play.google.com/store/apps/details?id=org.plantnet.

dataset to train four GoogLeNet's, one ResNet-152 and one ResNeXT. In Run 3, two additional GoogLeNet's and one ResNeXT were trained using a filtered version of the web dataset and images of the test set that received a probability higher than 0.98 in Run 1. The last and "winning" run *MarioTsaBerlin Run 4* finally combined all the 12 trained models.

**KDE TUT, Japan, 4 Runs, [4]** This participant introduced a modified version of the ResNet-50 model. Three of the intermediate convolutional layers used for downsampling were modified by changing the stride value from 2 to 1 and preceding it by max-pooling with a stride of two, to optimize the coverage of the inputs. Additionally, they switched the downsampling operation with the convolution for delaying the downsampling operation. This has been shown to improve performance by the authors of the ResNet architecture themselves. During the training they used data augmentation based on random crops, rotations and optional horizontal flipping. Test images were also augmented through a single flip operation and the resulting predictions averaged. Since the original ResNet-50 architecture was modified, no fine-tuning was used and the weights were learned from scratch starting with a big learning rate value of 0.1. The learning rates were multiplied by 0.1 twice, throughout the training process, over 100 epochs according to a schedule ratio 4:2:1 indicating the number of iterations using the same learning rate (limited to a total number of 350,000 iterations in the case of the big noisy dataset due to technical limitations). Run 1, 2, 3 were trained respectively on the trusted dataset, noisy dataset, and both datasets. The final run 4 is a combination of the outputs of the 3 runs.

**CMP, Czech Republic, 4 Runs, [2]** This participant based his work on the Inception-ResNet-v2 architecture [29] which introduces inception modules with residual connections. An additional maxout fully-connected layer with batch normalization was added on top of the network, before the classification fully-connected layer. Hard bootstrapping was used for training with noisy labels. A total of 17 models were trained using different training strategies such as: with or without maxout, with or without pre-training on ImageNet, with or without bootstrapping, with and without filtering of the noisy web dataset. CMP Run 1 is the combination of all the 17 networks by averaging their results. CMP Run 3 is the combination of the eight networks that were trained on the trusted EOL data solely. CMP Run 2 and CMP Run 4 are post-processings of CMP Run 1 and CMP Run 3 aimed at compensating the asymmetry of class distributions between the test set and the training sets.

### 8.3.3 Results

We report in Fig. 8.1 the performance achieved by the 29 evaluated systems. The used evaluation metric is the Mean Reciprocal Rank (MRR), i.e. the mean of the inverse of the rank of the correct species in the predictions returned by the evaluated system.

The first main outcome of that experiment was that the identification performance of state-of-the-art machine learning systems is impressive (with a median MRR around 0.8 and a maximal MRR of 0.92 for the best evaluated system *Mario MNB Run 4*). A second important conclusion was that the best results were obtained by the systems that were trained on both the trusted and the noisy dataset. Nevertheless, the systems that were trained exclusively on the noisy data (KDE TUT Run 2 and UM Run 2) performed better than the ones using the trusted data solely. This demonstrates that crawling the web without any filtering is a very effective way of creating large-scale training sets of plant observations. It opens the door to the possibility of building even larger systems working at the scale of the world's flora (or at least on 100 K species). Regarding the machine learning methods used by the participants, it is noticeable that all evaluated systems were based on Convolutional Neural Networks (CNN) confirming definitively the supremacy of this kind of approach over previous methods. A wide variety of popular architectures were trained from scratch or fine-tuned from pre-trained weights on the popular ImageNet dataset: GoogLeNet [9] and its improved inception v2 [23] and v4 [29] versions, inception-resnet-v2 [29], ResNet-50 and ResNet-152 [22], ResNeXT [22], VGGNet [28] and even the older AlexNet [26]. Another noticeable conclusion was that the best results were obtained with ensemble classifiers. The best system Mario MNB Run 4, for instance, was based on the aggregation of 12 CNNs (7 GoogLeNet, 2 ResNet-152, 3 ResNeXT). The CMP team combined also numerous models, a total of 17 models for instance for the CMP Run 1 with various sub-training datasets and bagging

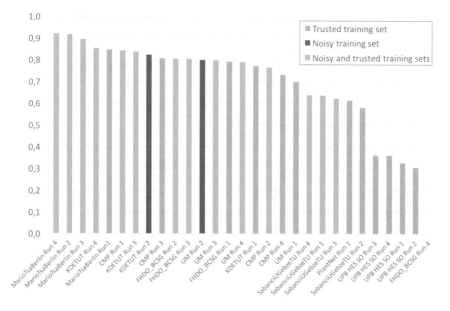

**Fig. 8.1** Performance achieved by the 29 systems evaluated within the plant identification challenge of LifeCLEF 2017

strategies, but all with the same inception-resnet-v2 architecture. Another key for succeeding the task was the use of data augmentation with usual transformations such as random cropping, horizontal flipping, rotation, for increasing the number of training samples and helping the CNNs to generalize better. Mario MNB team added two more interesting transformations, color saturation and lightness.

## 8.4 Human vs. Machine Experiment

The amazingly high performance of machine learning techniques measured within the LifeCLEF 2017 challenge raises several questions regarding automated species identification: Is there still a margin of progression? Are machine learning algorithms becoming as effective as human experts? What is the maximum reachable performance when using only images as the main source of information? As discussed above, a picture actually contains only a partial information about the observed plant and it is often not sufficient to determine the right species with certainty. Estimating this intrinsic uncertainty, thanks to human experts, is thus of crucial interest to answer the question of whether the problem is solved from a computer science perspective. Therefore, we conducted two experiments described in the two following subsections. The first one (Sect. 8.4.1) extends the results of the previous *Human vs. Machine experiment* that we conducted in 2014. It aims at measuring the progress that were made by automated identification systems since that time. The second experiment is based on a new testbed involving more challenging species and a panel of botanists with a much higher expertise on the targeted flora. It aims at answering the main questions asked in this paper. In the aim to start to response to these answers, we conducted several experiments with the some of the most state-of-the-art automated plant identification methods.

### *8.4.1 Progress Made Since 2014*

As discussed above, a first human vs. machine experiment [1] was conducted in 2014 based on 100 botanical observations that were identified by a panel of people with various expertise as well as by the systems evaluated within the LifeCLEF 2014 challenge. The 100 plants were selected at random from the whole set of observations of the PlantCLEF 2014 dataset [13]. This reduced test set was then shared with a large audience of potential volunteers composed of four target groups: **expert of the Flora** (highly skilled people such as taxonomists, expert botanists of the considered flora), **expert** (skilled people like botanists, naturalists, teachers, but not necessarily specialized on the considered Flora), **amateur** (people interested by plants in parallel of their professional activity and having a knowledge at different expertise levels), and **novice** (inexperienced users). The identification propositions were collected through a user interface presenting the 100 observations one by one (with one or several pictures of the different organs) and allowing the user to select up to three species for each observation thanks to a drop-down menu covering the 500 species of the PlantCLEF 2014 dataset. The most popular common names were also displayed in addition to the scientific name of the taxon to facilitate the participation of amateurs and novices. If the user didn't provide any species proposition for a given observation, the rank of the correct species was considered as infinite in the evaluation metric. We restricted the evaluation to the knowledge-based identification of plants, without any additional information or tools during the test. Concretely, the participants were not allowed to use external resources like field guides or Flora books. Among all contacted people, 20 of them finally accepted to participate: 1 expert of the French flora, 7 from the expert group, 7 from the amateur group, 5 from the novice group.

The performance of the 27 systems evaluated within LifeCLEF 2014 were computed on the same 100 observations than the ones identified by the human participants. To allow a fair comparison with human-powered identifications, the number of propositions was also limited to three (i.e. to the three species with the highest score for each test observation). To measure the progress since 2014, we did propose to the research groups who participated to the 2017-th edition of LifeCLEF to run their system on the same testbed. The three research groups who developed the best performing systems accepted to do so but only two of them (CMP and KDE TUT) were eligible for that experiment (the systems of Mario MNB were actually trained on a dataset that contained the 100 observations of the test set). Figure 8.2 reports the Mean Reciprocal Rank scores obtained by all human participants and all automated identification systems ("machines"). The description of the systems that were evaluated in 2014 ("Machine 2014") can be found in [13]). The description of the systems that were evaluated in 2017 ("Machine 2017") can be found in Sect. 8.3.2.

The main outcome of Fig. 8.2 is the impressive progress that was made by machines between 2014 and 2017. This progress is mostly due to the use of recent deep convolutional neural network architectures but also to the use of a much larger training data. Actually, the systems experimented in 2014 were trained on 60.962 images, while the systems experimented in 2017 were trained on respectively 256,287 pictures (EOL data) for CMP Run 3 and CMP Run 4, KDE TUT Run 1, and on 1.1 M pictures (EOL + Web) for the other ones. Interestingly, the fact that the 2017 systems were trained on 10 K species rather than 500 species did not affect their performance to much (this might even have increased their performance).

To conclude this first experiment with regard to our central question, one can notice that the quality of the identifications made by the best evaluated system is very close to the one of the only highly skilled botanist (qualified as "Expert of the flora" in Fig. 8.2). All other participants, including the botanists who were not directly specialists of the targeted flora, were outperformed by the five systems experimented in 2017.

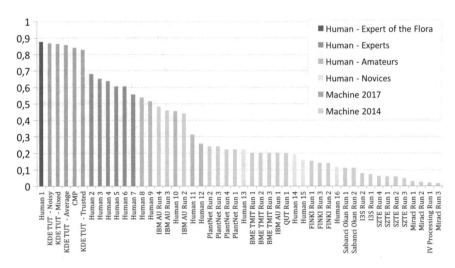

**Fig. 8.2** Identification performance of automated systems and humans of various expertise on the 2014-th test set

## 8.4.2   Experts vs. Machines Experiment (2017)

In the aim to evaluate more precisely the capacities of state-of-the-art plant identification systems compared to human expertise, we did set up a new evaluation with (1) a more difficult test set and (2), a group of highly skilled experts composed of the most renowned botanists of the considered flora.

### 8.4.2.1   Test Set Description

The new test set was created according to the following procedure. First, 125 plants were photographed between May and June 2017, a suitable period for the observation of flowers in Europe, in a botanical garden called the "Parc floral de Paris", and in a natural area located in the north of Montpellier city (southern part of France, close to the Mediterranean sea). The photos have been done with two smartphone models, an iPhone 5 and a Samsung S5 G930F, by a botanist and an amateur under his supervision. The selection of the species has been motivated by several criteria including (1) their membership to a difficult plant group (i.e. a group known as being the source of many confusions), (2) the availability of well developed specimens with well visible organs on the spot and (3), the diversity of the selected set of species in terms of taxonomy and morphology. About fifteen pictures of each specimen were acquired in order to cover all the informative parts of the plant. However, all pictures were not included in the final test set in order to deliberately hide a part of the information and increase the difficulty of the identification. Therefore, a random selection of only 1–5 pictures was operated for each specimen. In the end, a subset of 75 plants illustrated by a total of 216 images related to 33 families and 58 genera was selected. This test set is available online[5] under an open data license (CC0) in order to foster further evaluations by other research teams.

### 8.4.2.2  Experiment Description

The test set was sent to 20 expert botanists, working part-time or full-time as taxonomist, botanist, or research scientist specialist of the considered flora. Few of them were recognized as non-professional expert botanists. Most of them are or were involved (1) in the conception of renowned books or tools dedicated to the French flora (2) or in the study of large plant groups such as: Mediterranean flora [30]; Flora of ile-de-France [25]; Flora of cultivated fields [24]; author of the French national reference checklist [16]; author of the study of traits of Mediterranean species [27], publication on FloreNum,[6] etc. In addition to the test set, we provided to the experts an exhaustive list of 2567 possible species, which is basically the subpart of the 10,000 species used in PlantCLEF2017 related to the French flora exclusively. Regarding the difficulty of the task and contrary to the previous human vs. machine experiment done in 2014, each participant was allowed to use any external resource (book, herbarium material, computational tool, web app, etc.), excepted automated plant identification tools such as Pl@ntNet. For each plant, the experts were allowed to propose up to three species names ranked by decreasing confidence. Among the 20 contacted experts, 9 of them finally completed the task on time and returned their propositions.

In parallel, we did propose to the research groups who participated to the 2017-th edition of LifeCLEF to run their system on the same testbed than the one sent to the experts. The three research groups who developed the best performing systems accepted to do so and provided a total of 9 run files containing the species predictions of their systems with different configurations (see Sect. 8.3.2 for more details).

### 8.4.2.3  Results

Figure 8.3 displays the top-1 identification accuracy achieved by both the experts and the automated systems. Table 8.1 reports additional evaluation metrics namely the Mean Reciprocal Rank score (MRR), the top-2 accuracy and the top-3 accuracy. As a first noticeable outcome, none of the botanist correctly identified all observations. The top-1 accuracy of the experts is in the range 0.613–0.96. with a median value of 0.8. This illustrates the high difficulty of the task, especially when reminding that the experts were authorized to use any external resource to complete

---

[5]http://otmedia.lirmm.fr/LifeCLEF/mvsm2017/.

[6]http://www.florenum.fr/.

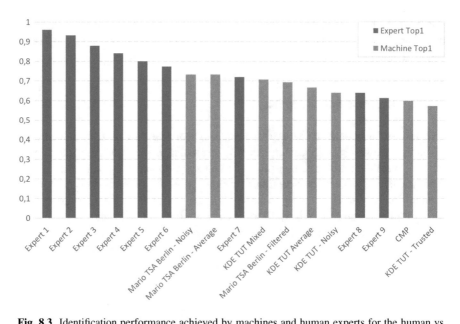

**Fig. 8.3** Identification performance achieved by machines and human experts for the human vs. machine 2017 experiments

**Table 8.1** Results of the human vs. machine 2017 experiments ordered by the top 1 accuracy

| Run | RunType | MRR | Top1 | Top2 | Top3 |
|---|---|---|---|---|---|
| Expert 1 | Human | 0.967 | 0.96 | 0.973 | 0.973 |
| Expert 2 | Human | 0.947 | 0.933 | 0.96 | 0.96 |
| Expert 3 | Human | 0.88 | 0.88 | 0.88 | 0.88 |
| Expert 4 | Human | 0.864 | 0.84 | 0.88 | 0.893 |
| Expert 5 | Human | 0.8 | 0.8 | 0.8 | 0.8 |
| Expert 6 | Human | 0.78 | 0.773 | 0.787 | 0.787 |
| Mario TSA Berlin—Noisy | Machine | 0.819 | 0.733 | 0.827 | 0.893 |
| Mario TSA Berlin—Average | Machine | 0.805 | 0.733 | 0.813 | 0.853 |
| Expert 7 | Human | 0.74 | 0.72 | 0.76 | 0.76 |
| KDE TUT Mixed | Machine | 0.786 | 0.707 | 0.8 | 0.827 |
| Mario TSA Berlin—Filtered | Machine | 0.751 | 0.693 | 0.747 | 0.787 |
| KDE TUT Average | Machine | 0.753 | 0.667 | 0.76 | 0.787 |
| Expert 8 | Human | 0.64 | 0.64 | 0.64 | 0.64 |
| KDE TUT—Noisy | Machine | 0.75 | 0.64 | 0.8 | 0.813 |
| Expert 9 | Human | 0.62 | 0.613 | 0.627 | 0.627 |
| CMP | Machine | 0.679 | 0.6 | 0.667 | 0.72 |
| KDE TUT—Trusted | Machine | 0.656 | 0.573 | 0.613 | 0.72 |
| Mario TSA Berlin—Trusted | Machine | 0.646 | 0.56 | 0.64 | 0.68 |

the task, Flora books in particular. It shows that a large part of the observations in the test set do not contain enough information to be surely identified when using classical identification keys. Only the four experts with an exceptional field expertise were able to correctly identify more than 80% of the observations.

Besides, Fig. 8.3 shows that the top-1 accuracy of the evaluated systems is in the range 0.56–0.733 with a median value of 0.66. This is globally lower than the experts but it is noticeable that the best systems were able to perform similarly or slightly better than three of the highly skilled participating experts. Moreover, if we look at the top-3 accuracy values provided in Table 8.1, we can see that the best evaluated system returned the correct species within its top-3 predictions for more than 89% of the test observations. Only the two best experts obtained a higher top-3 accuracy. This illustrates one of the strength of the automated identification systems. They can return an exhaustive ranked list of the most probable predictions over all species whereas this is a very difficult and painful task for human experts. Figure 8.4 displays the further top-K accuracy values as a function of K for all the evaluated systems. It shows that the performance of all systems continues to increase significantly for values of K higher than 3 and then becomes more stable for values of K in the range [20–50]. Interestingly, the best system reaches a top-11 accuracy of 0.973%, i.e. the same value of the top-1 accuracy of the best expert, and a 100% top-K accuracy for $K = 39$. In view of the thousands of species in the whole check list, it is likely that such a system would be very useful even for the experts themselves. By providing an exhaustive short list of all the possible species, it would help them to not exclude any candidate species that they might have missed otherwise.

To further understand the limitations and the margin of progress of the evaluated identification systems, we did analyze more deeper which of the 75 test observations were correctly identified or missed compared to the expert's propositions. The main outcome of that analysis is that the automated systems perform as well as experts for about 86% of the observations, i.e. for 65 of the 75 test observations, the best evaluated system ranked the right species at a lower or equal rank than the best expert. Among the ten remaining observations, six were correctly identified in the top-3 predictions of the best system and nine in the top-5. Figure 8.5 displays 3 of the most difficult observations for the machines, i.e. the ones that were not identified by any system within the top-3 predictions. It is likely that the cause of the identification failure differs from an observation to another one. For the observation n°74, for instance, it is likely that the main cause of failure is a mismatch between the training data and the test sample. Actually, the training samples of that species usually contain visible flowers whereas only the leaves are visible in the test sample. For the observation n°29, it is more likely that the failure is due to the intrinsic difficulty of the *Carex* genus within which many species are very similar visually. Most of the proposals in machine runs are nevertheless under the *Carex* genus. For observation n°43, the fact that most of images were not focused on a single leaf but dedicated to the illustration of the whole plant, which has a common aspect of a tuft of leaves, is probably at the origin of the misidentification. The small size of the discriminant organs and the cluttered background in the test sample makes the identification even more difficult.

| Id | Species | Images |
|----|---------|--------|
| 74 | Lathyrus vernus (L.) Bernh | |
| 29 | Carex distans L. | |
| 43 | Apium graveolens L. | |

**Fig. 8.4** Examples of observations well identified by experts but missed by the automated identification systems

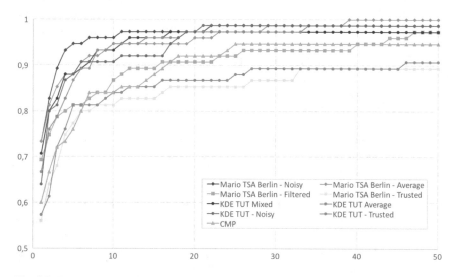

**Fig. 8.5** Top-K accuracy of the evaluated system as a function of K

## 8.5   Conclusion and Perspectives

The goal of this paper was to answer the question of whether automated plant identification systems still have a margin of progression or if they already perform as well as experts for identifying plants in images. Our study first shows that identifying plants from images solely is a difficult task, even for some of the highly skilled specialists who accepted to participate to the experiment. This confirms that pictures of plants only contain partial information and that it is often not sufficient to determine the right species with certainty. Regarding the performance of the machine learning algorithms, our study shows that there is still a margin of progression but that it is becoming tighter and tighter. Indeed, the evaluated systems were able to correctly identify as many plants as three of the experts whereas all of them were specialists of the considered flora. The best system was able to correctly classify 73.3% of the test samples including some belonging to very difficult taxonomic groups. This performance is still far from the best expert who correctly identified 96.7% of the test samples, however, as shown in our study, a strength of the automated systems is that they can return instantaneously an exhaustive list of all the possible species whereas this is a very difficult task for humans. We believe that this already makes them highly powerful tools for modern botany. Indeed, classical field guides or identification keys are much more difficult to handle and they require much more time to achieve a similar result. Furthermore, the performance of automated systems will continue to improve in the following years thanks to the quick progress of deep learning technologies. It is likely that systems capable of identifying the entire world's flora will appear in the next few years. The real question now is how to integrate them in pedagogical tools that could be used in teaching programs effectively and in a sustainable way. They have the potential to become essential tools for teachers and students, but they should not replace an in-depth understanding of botany.

**Acknowledgements** Most of the work conducted in this paper was funded by the Floris'Tic initiative, especially for the support of the organization of the PlantCLEF challenge. Milan Šulc was supported by CTU student grant SGS17/185/OHK3/3T/13. Valéry Malécot was supported by ANR ReVeRIES (ref: ANR-15-CE38-0004-01). Authors would like to thank the botanists who accepted to participate to this challenge : Benoit Bock (PhotoFlora), Nicolas Georges (Cerema), Arne Saatkamp (Aix Marseille Université, IMBE), François-Jean Rousselot, and Christophe Girod.

## References

1. Bonnet, P., Joly, A., Goëau, H., Champ, J., Vignau, C., Molino, J. F., Barthélémy Daniel & Boujemaa, N. (2016). Plant identification: man vs. machine. Multimedia Tools and Applications, 75(3), 1647–1665.
2. Sulc, M., & Matas, J. (2017). Learning with noisy and trusted labels for fine-grained plant recognition. Working Notes of CLEF, 2017.

3. Ludwig, A. R., Piorek, H., Kelch, A. H., Rex, D., Koitka, S., & Friedrich, C. M. (2017). Improving model performance for plant image classification with filtered noisy images. Working Notes of CLEF, 2017.
4. Hang, S. T., & Aono, M. (2017). Residual network with delayed max pooling for very large scale plant identification. Working Notes of CLEF, 2017.
5. Lasseck, M. (2017). Image-based plant species identification with deep convolutional neural networks. Working Notes of CLEF, 2017.
6. Atito, S., Yanikoglu, B., & Aptoula, E. Plant Identification with Large Number of Species: SabanciU-GebzeTU System in PlantCLEF 2017.
7. Lee, S. H., Chang, Y. L., & Chan, C. S. (2017). Lifeclef 2017 plant identification challenge: Classifying plants using generic-organ correlation features. Working Notes of CLEF, 2017.
8. Toma, A., Stefan, L. D., & Ionescu, B. (2017). Upb hes so@ plantclef 2017: Automatic plant image identification using transfer learning via convolutional neural networks. Working Notes of CLEF, 2017.
9. Szegedy, C., Liu, W., Jia, Y., Sermanet, P., Reed, S., Anguelov, D., ... & Rabinovich, A. (2015). Going deeper with convolutions. In Proceedings of the IEEE conference on computer vision and pattern recognition (pp. 1–9).
10. Goëau, H., Bonnet, P., & Joly, A. (2017). Plant identification based on noisy web data: the amazing performance of deep learning (lifeclef 2017). CEUR Workshop Proceedings.
11. Krause, J., Sapp, B., Howard, A., Zhou, H., Toshev, A., Duerig, T., ... & Fei-Fei, L. (2016, October). The unreasonable effectiveness of noisy data for fine-grained recognition. In European Conference on Computer Vision (pp. 301–320). Springer International Publishing.
12. Goëau, H., Bonnet, P., & Joly, A. (2015). LifeCLEF Plant Identification Task 2015. CEUR Workshop Proceedings.
13. Goëau, H., Joly, A., Bonnet, P., Selmi, S., Molino, J. F., Barthélémy, D., & Boujemaa, N. (2014). Lifeclef plant identification task 2014. In CLEF2014 Working Notes. Working Notes for CLEF 2014 Conference, Sheffield, UK, September 15–18, 2014 (pp. 598–615). CEUR-WS.
14. Deng, J., Dong, W., Socher, R., Li, L. J., Li, K., & Fei-Fei, L. (2009, June). Imagenet: A large-scale hierarchical image database. In Computer Vision and Pattern Recognition, 2009. CVPR 2009. IEEE Conference on (pp. 248–255). IEEE.
15. Farnsworth, E. J., Chu, M., Kress, W. J., Neill, A. K., Best, J. H., Pickering, J., ... & Ellison, A. M. (2013). Next-generation field guides. BioScience, 63(11), 891–899.
16. Bock B. (2014) Référentiel des trachéophytes de France métropolitaine réalisé dans le cadre d'une convention entre le Ministère chargé de l'Ecologie, le MNHN, la FCBN et Tela Botanica. Editeur Tela Botanica. Version 2.01 du 14 février 2014.
17. Gaston, K. J., & O'Neill, M. A. (2004). Automated species identification: why not?. Philosophical Transactions of the Royal Society of London B: Biological Sciences, 359(1444), 655–667.
18. Goëau, H., Bonnet, P., & Joly, A. (2016). Plant identification in an open-world (lifeclef 2016). In Working Notes of CLEF 2016-Conference and Labs of the Evaluation forum, évora, Portugal, 5–8 September, 2016. (pp. 428–439).
19. Goëau, H., Bonnet, P., & Joly, A., Yahiaoui I., Barthelemy D., Boujemaa N., Molino J.-f. (2012). The ImageCLEF 2012 Plant Identification Task. CEUR Workshop Proceedings.
20. Goëau, H., Joly, A., Bonnet, P., Bakic, V., Barthélémy, D., Boujemaa, N., & Molino, J. F. (2013, October). The imageCLEF plant identification task 2013. In Proceedings of the 2nd ACM international workshop on Multimedia analysis for ecological data (pp. 23–28). ACM.
21. Goëau, H., Bonnet, P., Joly, A., Boujemaa, N., Barthelemy, D., Molino, J. F., ... & Picard, M. (2011, September). The ImageCLEF 2011 plant images classi cation task. In ImageCLEF 2011.
22. He, K., Zhang, X., Ren, S., & Sun, J. (2016). Deep residual learning for image recognition. In Proceedings of the IEEE conference on computer vision and pattern recognition (pp. 770–778).
23. Ioffe, S., & Szegedy, C. (2015, June). Batch normalization: Accelerating deep network training by reducing internal covariate shift. In International Conference on Machine Learning (pp. 448–456).
24. Jauzein P. (1995). Flore des champs cultivés. Num.3912, Editions Quae.

25. Jauzein P., Nawrot O. (2013). Flore d'Ile-de-France: clés de détermination, taxonomie, statuts, Editions Quae.
26. Krizhevsky, A., Sutskever, I., & Hinton, G. E. (2012). Imagenet classification with deep convolutional neural networks. In Advances in neural information processing systems (pp. 1097–1105).
27. Saatkamp, A., Affre, L., Dutoit, T., & Poschlod, P. (2011). Germination traits explain soil seed persistence across species: the case of Mediterranean annual plants in cereal fields. Annals of botany, 107(3), 415–426.
28. Simonyan, K., & Zisserman, A. (2014). Very deep convolutional networks for large-scale image recognition. arXiv preprint arXiv:1409.1556.
29. Szegedy, C., Ioffe, S., Vanhoucke, V., & Alemi, A. A. (2017). Inception-v4, Inception-ResNet and the Impact of Residual Connections on Learning. In AAAI (pp. 4278–4284).
30. Tison, J. M., Jauzein, P., Michaud, H., & Michaud, H. (2014). Flore de la France méditerranéenne continentale. Turriers: Naturalia publications.

# Chapter 9
# Automated Identification of Herbarium Specimens at Different Taxonomic Levels

Jose Carranza-Rojas, Alexis Joly, Hervé Goëau, Erick Mata-Montero, and Pierre Bonnet

**Abstract** The estimated number of flowering plant species on Earth is around 400,000. In order to classify all known species via automated image-based approaches, current datasets of plant images will have to become considerably larger. To achieve this, some authors have explored the possibility of using herbarium sheet images. As the plant datasets grow and start reaching the tens of thousands of classes, unbalanced datasets become a hard problem. This causes models to be inaccurate for certain species due to intra- and inter-specific similarities. Additionally, automatic plant identification is intrinsically hierarchical. In order to tackle this problem of unbalanced datasets, we need ways to classify and calculate the loss of the model by taking into account the taxonomy, for example, by grouping species at higher taxon levels. In this research we compare several architectures for automatic plant identification, taking into account the plant taxonomy to classify not only at the species level, but also at higher levels, such as genus and family.

J. Carranza-Rojas (✉) · E. Mata-Montero
School of Computing, Costa Rica Institute of Technology, Cartago, Costa Rica
e-mail: jcarranza@itcr.ac.cr; emata@itcr.ac.cr

A. Joly
Inria ZENITH Team, Montpellier, France
e-mail: alexis.joly@inria.fr

H. Goëau · P. Bonnet
CIRAD, UMR AMAP, Montpellier, France

AMAP, Univ Montpellier, CIRAD, CNRS, INRA, IRD, Montpellier, France
e-mail: herve.goeau@cirad.fr; pierre.bonnet@cirad.fr

© Springer International Publishing AG, part of Springer Nature 2018
A. Joly et al. (eds.), *Multimedia Tools and Applications for Environmental & Biodiversity Informatics*, Multimedia Systems and Applications,
https://doi.org/10.1007/978-3-319-76445-0_9

## 9.1   Introduction

In general, Deep Learning classification has focused mostly on flat classification, i.e., hierarchies and knowledge associated with higher levels are usually not taken into account. However, in the biological domain, the approach traditionally followed by taxonomists is intrinsically hierarchical. Single-access and multiple-access identification keys are an example of such an approach [3]. They are used to identify organisms mostly at the species level but sometimes at the genus and family levels too. To our knowledge, most of the research on image-based automated plant identifications classify plant images into species and do not exploit knowledge about other taxonomic levels.

Very few studies also have attempted to use herbarium images for plant identification. With new deep learning methods, large datasets of herbarium images such as those published by iDigBio[1] [1, 2], which comprises millions of images of thousands of species from around the globe, become very useful. These datasets are suitable for deep learning approaches and include as metadata all levels of the associated taxonomy. In [4] a GoogleNet model with modifications is used to classify species from the Mediterranean region and Costa Rica. It shows promising results in terms of accuracy when training and testing with herbarium sheet images, as well as when doing transfer learning from the Mediterranean region to Costa Rica. However, classifications are conducted only at the species level and do not use additional knowledge related to other taxonomic levels.

Herbaria normally hold many samples that have not been identified at the species level [5] but they make an effort to at least have them identified at the genus or family level. It is therefore important to help streamline the identification process with tools that support identifications at multiple levels (probably with different levels of accuracy).

One of the biggest issues in plant identification is the lack of balanced datasets. At the species level, most available datasets are unbalanced due to taxonomically uneven nature of sample collection processes [3]. So, an expected intuition in this domain is to exploit higher levels of the taxonomy in order to have more images of a single class and use that knowledge to help the classification at lower levels of the taxonomy, such as the species at the bottom. In other words, the unbalanced dataset issue could be tackled by using a class hierarchy and doing classifications from top to bottom.

In this work we compare several deep learning architectures to do herbarium specimen identification at not only species level, but also other taxonomic levels such as genus and family. We explore architectures that do several taxonomic level classifications at the same time by sharing parameters, as well as separated flat classifiers, independent from each other.

---

[1]https://www.idigbio.org/.

The rest of this manuscript is organized as follows: Sect. 9.2 presents relevant related work. Sections 9.3 and 9.4 cover methodological aspects and experiment design, respectively. Section 9.5 describes the results obtained. Section 9.6 presents the conclusions and, finally, Sect. 9.7 summarizes future work.

## 9.2  Related Work

PlantCLEF is the largest and best known plant identification challenge [6]. It has helped to create bigger datasets each year as well as allowed participants to gradually improve the techniques (mostly deep learning based models) to achieve better accuracy. So far, PlantCLEF has focused on species level identifications only.

The same situation happens with apps for automated image-based plant identification such as LeafSnap [7] and Pl@ntNet [8]. These apps are also focused on classification only at the species level; however, it would be useful in cases where the accuracy is low, to have predictions at other taxonomic levels such as genus and family.

Very few studies have tackled the problem of hierarchical classification. Silla et al. [9] present a very comprehensive survey about different techniques used for hierarchical classification. Wu et al. [10] discuss how there are no even proper standards to evaluate hierarchical classification systems, and use Naive Bayes approach on text data. Both studies are focused on traditional machine learning, not deep learning.

Shahbaba et al. [11] create a new method using a Bayesian form of the softmax function, adding a prior that introduces correlations between the parameters of nearby classes of the hierarchy. This approach was also developed for traditional machine learning and not deep learning. Nevertheless it could be easily adjustable to deep learning. This approach is also useful when there is a prior knowledge of the class hierarchy.

Yan et al. [12] create a new architecture named Hierarchical Deep CNN (HD-CNN), which uses two levels of classification. The first level is more general and then the second level is composed by several smaller classifiers per each class. This means that the amount of classifiers grows after the first classification. Furthermore, an error during the first classification will lead to error propagation to the second layer of classifiers.

There have been a lot of studies where the hierarchy is learned via unsupervised learning. In this paper, we focus on an already defined hierarchy which is a plant taxonomy. It is the result of decades if not centuries of work in the field of taxonomy. Based on that result, we don't calculate automatically the class hierarchy.

In particular, to our knowledge, no plant identification system or study has been proposed that actually exploits the class hierarchy using the plant taxonomy. In [13] authors do analyze the accuracy per species but also per genera and families, to see which species are better identified and also to evaluate if the amount of images per species has a direct impact over the accuracy obtained per class. They conclude that species with around 100 images are very well identified with some exceptions. Indeed, a few species are very well identified even with a very small number of images. They also provide accuracy per genus and family.

## 9.3 Methodology

Previous work in [4] has tackled the problem of using a big dataset with herbarium images for automatic plant identification. We describe the herbarium dataset taken from this study, and used for this research. We have also added information about genera and families, beyond species, in order to test the hierarchical architectures.

### 9.3.1 Datasets

Herbarium data used in the experiments comes from the iDigBio portal, which aggregates and gives access to millions of images for research purposes. As illustrated in Fig. 9.1, typical herbarium sheets result in a significantly affected visual representation of the plant, with a typical monotonous aspect of brown and dark green content and a modified shape of the leaves, fruits or flowers due to the drying process and aging. Moreover, the sheets are surrounded by handwritten/typewritten labels, bar codes, institutional stamps and reference colour bar patterns for the most recent ones.

Additionally, ImageNet weights were used to pre-train the deep learning model. We only used the weights of a pre-trained model on ImageNet, not the dataset itself. Details of the used datasets are presented below:

- Herbaria1K (H1K): this dataset covers 1191 species, 918 of which are included in the 1000 species of the PlantCLEF dataset from 2015 [14]. Obtained through iDigBio, the dataset contains 202,445 images for training and 51,288 images for testing. All images have been resized to a width of 1024 pixels and their height proportionally, given the huge resolutions used in herbarium images. This is an unbalanced dataset, as explained in next sections of this manuscript. In terms of genera, it contains a total of 498 genera, and regarding families it has a total of 124 families.
- ImageNet is the most widely used dataset by the machine learning research community. It contains 1000 generalist classes and more than a million images [15]. It is the *de facto* standard for pre-training deep learning models. We use only the weights of a trained model with this dataset for transfer learning proposes (Table 9.1).

**Fig. 9.1** Arctium minus
*(hill) bernh.* herbarium sheet
sample taken from the
Herbarium of Paris

**Table 9.1** Datasets used in this research

| Name | Acronym | Source | Type | # of images | # of species/ classes |
|------|---------|--------|------|-------------|------------------------|
| Herbarium1K | H1K | iDigBio | Herbarium sheets | 253,733 | 1191 |
| ImageNet | I | ImageNet challenge | Generic images | 1 M | 1000 |

## 9.3.2   Unbalanced Dataset

Figure 9.2 shows how unbalanced the H1K dataset is. According to the work in [13], the H1K dataset allows high identification rates with their deep learning model after 100 images per species. As shown in the figure, around 60% of the species have more than 100 images, and 40% less than that. Some species have lots of images, for example 324 species have more than 300 images, but in contrast, 311 species have less than 11 images in total for both training and testing.

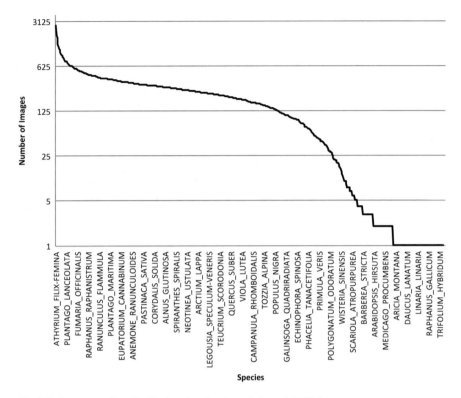

**Fig. 9.2** Image per class distribution showing the unbalanced H1K dataset

## 9.3.3 Architectures

The basis architecture used in our study is an extension of the GoogleNet architecture with batch norm [16], as used in [4] for plant identification on herbarium specimens. The main difference is at the last fully connected layer. Table 9.2 shows the modified GoogleNet network used in this research, taken from [4]. The network was implemented in Lasagne [17], using Theano [18].

### 9.3.3.1  Baseline: Flat Classification Model (FCM)

In order to evaluate the performance of adding hierarchies to the architecture classification, a first base line is set based on a FCM. Since we are classifying not only species but also genera and families, the flat approach requires three different instances of the same model, with different number of outputs on the last dense layer and softmax, according to the dataset label size for each taxonomic level. Figure 9.3 shows the three main building blocks that will be used on the next

**Table 9.2**  GoogleNet architecture modified with Batch Normalization, taken from [4]

| Type | Patch size/stride | Output size | Depth | Params | Ops |
|---|---|---|---|---|---|
| Convolution | $7 \times 7/2$ | $112 \times 112 \times 64$ | 1 | 2.7 K | 34 M |
| Max pool | $3 \times 3/2$ | $56 \times 56 \times 64$ | 0 | | |
| Batch norm | | $56 \times 56 \times 64$ | 0 | | |
| LRN | | $56 \times 56 \times 64$ | 0 | | |
| Convolution | $3 \times 3/1$ | $56 \times 56 \times 192$ | 2 | 112 K | 360 M |
| Max pool | $3 \times 3/2$ | $28 \times 28 \times 192$ | 0 | | |
| Batch norm | | $28 \times 28 \times 192$ | 0 | | |
| LRN | | $28 \times 28 \times 192$ | 0 | | |
| Inception (3a) | | $28 \times 28 \times 256$ | 2 | 159 K | 128 M |
| Inception (3b) | | $28 \times 28 \times 480$ | 2 | 380 K | 304 M |
| Max pool | $3 \times 3/2$ | $14 \times 14 \times 480$ | 0 | | |
| Batch norm | | $14 \times 14 \times 480$ | 0 | | |
| Inception (4a) | | $14 \times 14 \times 512$ | 2 | 364 K | 73 M |
| Inception (4b) | | $14 \times 14 \times 512$ | 2 | 437 K | 88 M |
| Inception (4c) | | $14 \times 14 \times 512$ | 2 | 463 K | 100 M |
| Inception (4d) | | $14 \times 14 \times 528$ | 2 | 580 K | 119 M |
| Inception (4e) | | $14 \times 14 \times 832$ | 2 | 840 K | 170 M |
| Max pool | $3 \times 3/2$ | $7 \times 7 \times 832$ | 0 | | |
| Batch norm | | $7 \times 7 \times 832$ | 0 | | |
| Inception (5a) | | $7 \times 7 \times 832$ | 2 | 1072 K | 54 M |
| Inception (5b) | | $7 \times 7 \times 1024$ | 2 | 1388 K | 71 M |
| Avg pool | $7 \times 7/1$ | $1 \times 1 \times 1024$ | 0 | | |
| Batch norm | | $1 \times 1 \times 1024$ | 0 | | |
| Linear | | $1 \times 1 \times 10,000$ | 1 | 1000 K | 1 M |
| Softmax | | $1 \times 1 \times 10,000$ | 0 | | |

**Fig. 9.3**  Representation of some building blocks of the different architectures

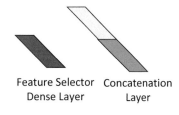

Feature Selector        Concatenation
Dense Layer               Layer

Fully Connected
Layer
With Softmax

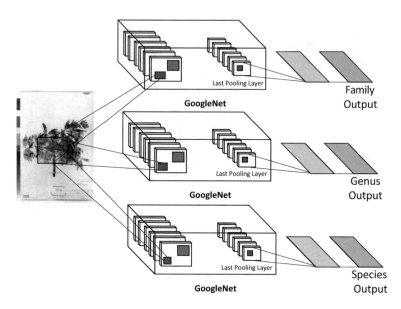

**Fig. 9.4** Separated Flat Classification Model (FCM) for species, genera and family

sections with information about the models. For species we have a total of 1191 outputs, for genera 498 and for families 124. These output sizes are the same across all architectures.

Figure 9.4 shows how the flat model looks like. The model is basically a GoogleNet [19] model, modified with Parametric REctified Linear Unit (PRELU) and batch normalization for faster convergence. A total of three different flat models were deployed: one for species, one for genera, and one for families. The three models are completely independent and do not share any parameters. They also have their own training and parameter update process.

### 9.3.3.2 Multi-Task Classification Model (MCM)

Another approach to calculate accuracy at different taxonomic levels is by using a model where the different classifiers share the same deep network. MCM implements one classifier per taxonomic level, in this case three classifiers, one for species, one for genera and one for families. However, each classifier is connected to the last pooling layer of the GoogleNet model, allowing to do three classifications at the same time and sharing the same parameters of the model instead of having three separate models with their own parameters. The intuition behind is that the network will learn features from the three taxonomic levels at the same time. Figure 9.5 shows how a single main GoogleNet model is shared between three different classifiers, each one assigned to classifying a different taxonomic level. This model is inspired by the work of [20], where the authors identify multi-digit numbers from houses, using a classifier per digit.

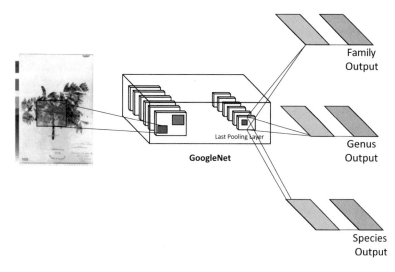

**Fig. 9.5** A Multi-Task Classification Model (MCM) for species, genera and family. Parameters are shared between the three taxonomic levels, similar to the work of [20] for multi-digit identification of house numbers

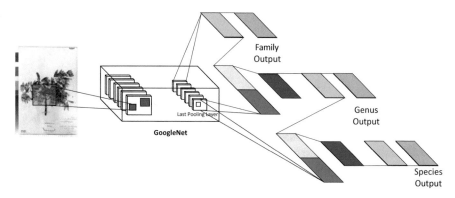

**Fig. 9.6** TaxonNet used to identify species, genera and family. The architecture allows to take into account previous classification of another taxonomic level for the next one

### 9.3.3.3 TaxonNet: Hierarchical Classification

We present the following architecture that attempts to capture features at several levels of the plant taxonomy. We call this architecture TaxonNet, as it takes into account several levels of the plant taxonomy as the hierarchy, and uses knowledge of the previous taxonomic level classification for the next one, as shown in Fig. 9.6.

We modified the GoogleNet model in the following fashion: the last fully connected layer which was used normally for a flat species classification is now used for the higher taxonomic level, in this case the Family. The loss of this fully connected layer will be calculated based on family labels of each image. Just before the softmax, the feature vector of the family fully connected layer output is concatenated with the last pooling layer feature vector. The idea behind this, is to add a new fully connected layer for the genus, which will base its computations on both the family fully connected feature vector, and the raw feature vector coming from the Convolutional Neural Network (CNN). In other words, the features learned to recognize the family will be used as a context to learn new complementary features at the genus level. Finally, we apply the same concept with the species: we add a new fully connected layer for species, which takes as input the concatenation of the genus fully connected layer output plus the last pooling layer feature vector from the CNN. In all cases, there is a middle feature selector layer in red, as shown in Fig. 9.3, which allows the model to learn which features to take into account either from the original GoogleNet or from the previous taxonomic classification.

It is important to notice that the design allows the model to make mistakes at higher levels of the taxonomy, such as family or genus, but it can have good accuracy at the species level since it handles the raw feature vectors coming from the CNN. In other words, an error at higher levels of the taxonomy won't necessarily cause an error propagation to lower levels. It also allows to do classification at all taxonomic levels, thus, each one has its own loss which is back-propagated to the whole network. Our intuition is that the whole network learns features at all taxonomic levels, instead of having several complete CNN for each level. This allows to share parameters between levels, for a smaller network.

## 9.4 Experiments

By using the previous explained models we ran several experiments to measure the effect of taking into account different taxonomic levels for the classification.

In all cases the used learning rate was 0.0075 and weight decay of 0.0002. The total number of training iterations was 6300 with a mini-batch size of 32 images, with 5 epochs. The number of validation iterations was 1500, the same as for testing iterations.

### 9.4.1 Baseline Experiments: Flat Classification Model (FCM)

The first experiments are based on running the separated models for species, genus and family without sharing any type of parameters. This is considered as the baseline, as there are no hierarchical characteristics at all, but just three models completely independent from each other.

## 9.4.2   Architecture Comparison Experiment

This experiment consists on a comparison of the different architectures at the different taxonomic levels. The experiment compares the MCM approach, where parameters are shared between the different classifiers, with the intention to see how the accuracy and loss behaves as the model is trained, compared to separated model per taxonomic level. The TaxonNet architecture is also compared with the separated models and the MCM aproach.

## 9.5   Results

### 9.5.1   FCM Baseline Results

First experiments consisted on running 3 separated FCM models to explore the loss and accuracy behavior at each taxonomic level. We consider this as the baseline results, as they are flat classifiers that do not share any hierarchical characteristics in terms of classification.

The results for all the FCM are shown in Figs. 9.7, 9.8 and 9.9. In particular, FCM for species gets Top-1 63.02% and Top-5 is about 82.93% as shown in Fig. 9.7. In case of the FCM for genus the accuracy goes up to Top-1 of 70.51% and Top-5 of 87.85%, as shown by Fig. 9.8. For the family, Fig. 9.9 shows the best results for both Top-1 and Top-5, with 75.55% and 93.43% respectively.

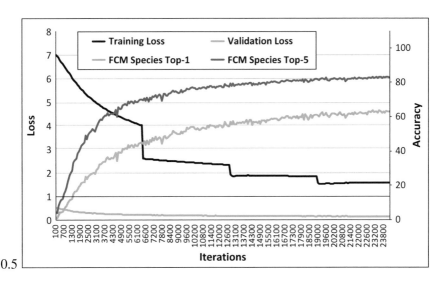

**Fig. 9.7**  FCM for species showing Top-1 and Top-5 accuracy and losses

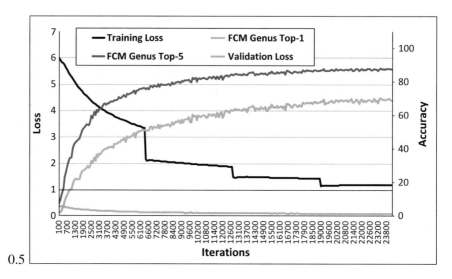

**Fig. 9.8** FCM for genera showing Top-1 and Top-5 accuracy and losses

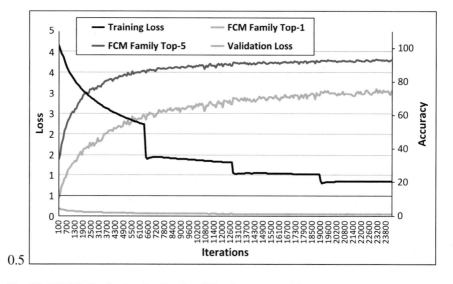

**Fig. 9.9** FCM for family showing Top-1 and Top-5 accuracy and losses

It is important to notice that both genus and family show an improvement compared to the species. This makes sense as genus and family have more images per class and also both models have less classes, 498 for genus and 124 for families. This compensates the difficulty of having a higher intra-class variability at that levels.

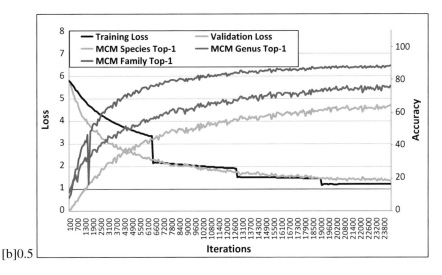

[b]0.5

**Fig. 9.10**  MCM results for species, genus and family on Top-1 accuracy

## 9.5.2   *Architecture Comparison Results*

### 9.5.2.1   MCM Top-1 and Top-5 Behavior

In case of the MCM architecture, for Top-1 accuracy the results show 64.32% for species, 75.95% for genus, and for family 88.17%, as shown by Fig. 9.10. The parameter sharing allows the model to predict the family with a very high accuracy. In case of Top-5 accuracy, MCM results in 71.66% for species, 83.23% for genus, and 92.99% for family. Family classification is again the best among the three, as shown by Fig. 9.11.

### 9.5.2.2   TaxonNet Top-1 and Top-5 Behavior

In Fig. 9.12, TaxonNet architecture shows for Top-1 accuracy 62.39%, 76.23%, 86.92% for species, genus and family, respectively. Again, similarly to MCM, the parameter sharing allows the model to predict the genus and family with a very high accuracy. For Top-5 accuracy, as shown by Fig. 9.13, TaxonNet results in 70.20%, 82.36% and 92.80% for species, genus and family, respectively, again being the family classification the best among the three.

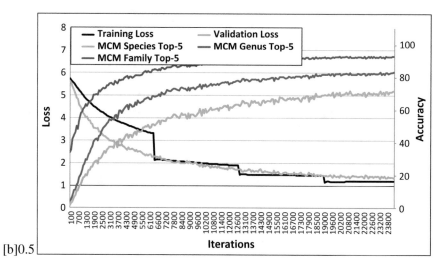

**Fig. 9.11** MCM results for species, genus and family on Top-5 accuracy

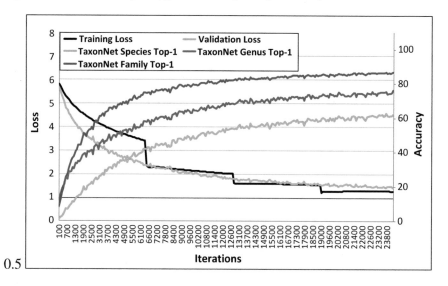

**Fig. 9.12** TaxonNet results for species, genus and family on Top-1 accuracy

### 9.5.2.3   Architecture Comparisons

Our results demonstrate that for species, Top-1 accuracy is 63.02%, 64.32%, 62.39% for FCM, MCM and TaxonNet, respectively, showing the best results on the MCM architecture by a margin of 1% approximately.

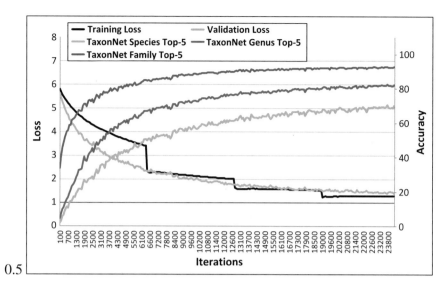

**Fig. 9.13** TaxonNet results for species, genus and family on Top-5 accuracy

Regarding genus, our result shows a Top-1 accuracy is 70.51%, 75.95%, 76.23% for FCM, MCM and TaxonNet, respectively. In this case, the degradation of the flat classifier for the genus is improved significantly by both hierarchical architectures, with the TaxonNet being the best one.

Finally, for family, our result shows a Top-1 accuracy is 75.55%, 88.17%, and 86.92% for FCM, MCM and TaxonNet. Here the improvement by both hierarchical architectures is very important compared to the flat classifier on Top-1.

## 9.6   Conclusions

The best accuracy results for species and genus are provided by the independent Flat Classification Model (FCM), but at the cost of three times more GPU consumption as well as three times more parameters. In case of the family, both the Multi-Task Classification Model (MCM) and TaxonNet architectures provide similar results to the flat model.

In general, whatever architecture is used, the classification accuracy increases significantly with the taxonomic level and reaches high classification accuracy at the family level whereas such groups are very heterogeneous visually as they include a lot of species.

The goal of this research was to study several architectures for automatic plant identification, taking into account the plant taxonomy to classify not only at the species level, but also at higher levels, in particular genus and family. In this regard, we introduced two architectures on top of the GoogleNet basis convolutional neural network: one multi-task classifier dedicated to do three classifications at the same time, and one hierarchical architecture (called TaxonNet) aimed at capturing features at several levels of the plant taxonomy. Our experiments did show that the multi-task network as well as the hierarchical one allow considerable improvements compared to having separate flat classifiers, in particular for predicting the genus and the family.

## 9.7   Future Work

This work uses knowledge of higher levels of taxonomy for species classification, and allows to classify at higher levels of the taxonomy such as genus and family. However, it uses traditional fully connected layers with traditional cross entropy loss and softmax calculations. Next steps include exploiting the class hierarchy to calculate a different loss functions using Bayesian approaches of hierarchical softmax functions. Furthermore, hierarchical regularization terms could be defined to regularize the loss calculation using the class hierarchy. Interesting future experiments include understanding how using the taxonomy impacts the classification of new, unseen classes, at higher taxon levels. For instance, a species may not have been included during training but the genus related to that species may be, thus, allowing the system to provide an identification at that level could be of a strong interest. Additional architectures are also needed to be explored such as Long Short-Term Memory (LSTM) based architectures for the taxonomy.

**Acknowledgements** Special thanks to the Colaboratorio Nacional de Computación Avanzada (CCNA) in Costa Rica for sharing their Tesla K40-based cluster and providing technical support for this research. We also thank the Costa Rica Institute of Technology for the financial support for this research.

## References

1. Matsunaga, A., Thompson, A., Figueiredo, R. J., Germain-Aubrey, C. C., Collins, M., Beaman, R. S., ... & Fortes, J. A. (2013, October). A computational-and storage-cloud for integration of biodiversity collections. In eScience (eScience), 2013 IEEE 9th International Conference on (pp. 78–87). IEEE.
2. Page, L. M., MacFadden, B. J., Fortes, J. A., Soltis, P. S., & Riccardi, G. (2015). Digitization of biodiversity collections reveals biggest data on biodiversity. BioScience, 65(9), 841–842.
3. E. Mata-Montero and J. Carranza-Rojas, *Automated Plant Species Identification: Challenges and Opportunities.* Springer International Publishing, 2016, pp. 26–36.
4. J. Carranza-Rojas, H. Goeau, P. Bonnet, E. Mata-Montero, and A. Joly, "Going deeper in the automated identification of herbarium specimens," *BMC Evolutionary Biology*, vol. 17, no. 1, p. 181, Aug 2017.

5. D. P. Bebber, M. A. Carine, J. R. Wood, A. H. Wortley, D. J. Harris, G. T. Prance, G. Davidse, J. Paige, T. D. Pennington, N. K. Robson et al., "Herbaria are a major frontier for species discovery," *Proceedings of the National Academy of Sciences*, vol. 107, no. 51, pp. 22 169–22 171, 2010.
6. A. Joly, H. Goëau, H. Glotin, C. Spampinato, P. Bonnet, W.-P. Vellinga, J. Champ, R. Planqué, S. Palazzo, and H. Müller, *LifeCLEF 2016: Multimedia Life Species Identification Challenges*. Cham: Springer International Publishing, 2016, pp. 286–310.
7. N. Kumar, P. N. Belhumeur, A. Biswas, D. W. Jacobs, W. J. Kress, I. C. Lopez, and J. V. Soares, "Leafsnap: A computer vision system for automatic plant species identification," in *Computer Vision–ECCV 2012*. Springer, 2012, pp. 502–516.
8. A. Joly, P. Bonnet, H. Goëau, J. Barbe, S. Selmi, J. Champ, S. Dufour-Kowalski, A. Affouard, J. Carré, J.-F. Molino et al., "A look inside the pl@ntnet experience," *Multimedia Systems*, vol. 22, no. 6, pp. 751–766, 2016.
9. C. N. Silla and A. A. Freitas, "A survey of hierarchical classification across different application domains," *Data Min Knowl Disc*, vol. 22, pp. 31–72, 2011.
10. F. Wu, J. Zhang, and V. Honavar, "Learning classifiers using hierarchically structured class taxonomies," in *Proceedings of the 6th International Conference on Abstraction, Reformulation and Approximation*, ser. SARA'05. Berlin, Heidelberg: Springer-Verlag, 2005, pp. 313–320. [Online]. Available: http://dx.doi.org/10.1007/11527862_24
11. B. Shahbaba and R. M. Neal, "Improving classification when a class hierarchy is available using a hierarchy-based prior," *Bayesian Anal.*, vol. 2, no. 1, pp. 221–237, 03 2007. [Online]. Available: http://dx.doi.org/10.1214/07-BA209
12. Z. Yan, H. Zhang, R. Piramuthu, V. Jagadeesh, D. DeCoste, W. Di, and Y. Yu, "Hd-cnn: Hierarchical deep convolutional neural network for large scale visual recognition," in *ICCV'15: Proc. IEEE 15th International Conf. on Computer Vision*, 2015.
13. J. Carranza-Rojas, A. A.J.ăJoly, P. Bonnet, H.H.G. ăGoëau, and E. Mata-Montero, "Automated herbarium specimen identification using deep learning," *Biodiversity Information Science and Standards*, vol. 1, p. e20302, 2017.
14. H. Goëau, P. Bonnet, and A. Joly, "LifeCLEF Plant Identification Task 2015," in *CLEF: Conference and Labs of the Evaluation forum*, ser. CLEF2015 Working notes, CEUR-WS, Ed., vol. 1391, Toulouse, France, Sep. 2015. [Online]. Available: https://hal.inria.fr/hal-01182795
15. O. Russakovsky, J. Deng, H. Su, J. Krause, S. Satheesh, S. Ma, Z. Huang, A. Karpathy, A. Khosla, M. Bernstein, A. C. Berg, and L. Fei-Fei, "ImageNet Large Scale Visual Recognition Challenge," *International Journal of Computer Vision (IJCV)*, vol. 115, no. 3, pp. 211–252, 2015.
16. S. Ioffe and C. Szegedy, "Batch normalization: Accelerating deep network training by reducing internal covariate shift," *CoRR*, vol. abs/1502.03167, 2015. [Online]. Available: http://arxiv.org/abs/1502.03167
17. S. Dieleman, J. Schlüter, C. Raffel, E. Olson, S. K. Sønderby, D. Nouri et al., "Lasagne: First release." Aug. 2015. [Online]. Available: http://dx.doi.org/10.5281/zenodo.27878
18. Theano Development Team, "Theano: A Python framework for fast computation of mathematical expressions," *arXiv e-prints*, vol. abs/1605.02688, May 2016. [Online]. Available: http://arxiv.org/abs/1605.02688
19. C. Szegedy, W. Liu, Y. Jia, P. Sermanet, S. Reed, D. Anguelov, D. Erhan, V. Vanhoucke, and A. Rabinovich, "Going deeper with convolutions," *Proceedings of the IEEE Computer Society Conference on Computer Vision and Pattern Recognition*, vol. 07-12-June, pp. 1–9, 2015.
20. I. J. Goodfellow, Y. Bulatov, J. Ibarz, S. Arnoud, and V. Shet, "Multi-digit number recognition from street view imagery using deep convolutional neural networks," 2014. [Online]. Available: https://arxiv.org/pdf/1312.6082.pdf

# Chapter 10
# A Deep Learning Approach to Species Distribution Modelling

Christophe Botella, Alexis Joly, Pierre Bonnet, Pascal Monestiez, and François Munoz

**Abstract** Species distribution models (SDM) are widely used for ecological research and conservation purposes. Given a set of species occurrence, the aim is to infer its spatial distribution over a given territory. Because of the limited number of occurrences of specimens, this is usually achieved through environmental niche modeling approaches, i.e. by predicting the distribution in the geographic space on the basis of a mathematical representation of their known distribution in environmental space (= realized ecological niche). The environment is in most cases represented by climate data (such as temperature, and precipitation), but other variables such as soil type or land cover can also be used. In this paper, we propose a deep learning approach to the problem in order to improve the predictive effectiveness. Non-linear prediction models have been of interest for SDM for more

C. Botella (✉)
INRIA Sophia-Antipolis - ZENITH Team, LIRMM - UMR 5506 - CC 477, Montpellier, France

INRA, UMR AMAP, Montpellier, France

AMAP, Univ Montpellier, CIRAD, CNRS, INRA, IRD, Montpellier, France

BioSP, INRA, Site Agroparc, Avignon, France

A. Joly
Inria ZENITH Team, Montpellier, France
e-mail: alexis.joly@inria.fr

P. Bonnet
CIRAD, UMR AMAP, Montpellier, France

AMAP, Univ Montpellier, CIRAD, CNRS, INRA, IRD, Montpellier, France
e-mail: pierre.bonnet@cirad.fr

P. Monestiez
BioSP, INRA, Site Agroparc, Avignon, France
e-mail: pascal.monestiez@inra.fr

F. Munoz
Université Grenoble Alpes, Saint-Martin-d'Hères, France
e-mail: francois.munoz@cirad.fr

© Springer International Publishing AG, part of Springer Nature 2018
A. Joly et al. (eds.), *Multimedia Tools and Applications for Environmental & Biodiversity Informatics*, Multimedia Systems and Applications,
https://doi.org/10.1007/978-3-319-76445-0_10

169

than a decade but our study is the first one bringing empirical evidence that deep, convolutional and multilabel models might participate to resolve the limitations of SDM. Indeed, the main challenge is that the realized ecological niche is often very different from the theoretical fundamental niche, due to environment perturbation history, species propagation constraints and biotic interactions. Thus, the realized abundance in the environmental feature space can have a very irregular shape that can be difficult to capture with classical models. Deep neural networks on the other side, have been shown to be able to learn complex non-linear transformations in a wide variety of domains. Moreover, spatial patterns in environmental variables often contains useful information for species distribution but are usually not considered in classical models. Our study shows empirically how convolutional neural networks efficiently use this information and improve prediction performance.

## 10.1 Introduction

### 10.1.1 Context on Species Distribution Models

Species distribution models (SDM) have become increasingly important in the last few decades for the study of biodiversity, macro ecology, community ecology and the ecology of conservation. An accurate knowledge of the spatial distribution of species is actually of crucial importance for many concrete scenarios including the landscape management, the preservation of rare and/or endangered species, the surveillance of alien invasive species, the measurement of human impact or climate change on species, etc. Concretely, the goal of SDM is to infer the spatial distribution of a given species based on a set of geo-localized occurrences of that species (collected by naturalists, field ecologists, nature observers, citizen sciences project, etc.). However, it is usually not possible to learn that distribution directly from the spatial positions of the input occurrences. The two major problems are the limited number of occurrences and the bias of the sampling effort compared to the real underlying distribution. In a real-world dataset, the raw spatial distribution of the observations is actually highly correlated to the preference and habits of the observers and not only to the spatial distribution of the species. Another difficulty is that in most cases, we only have access to presence data but not to absence data. In other words, occurrences inform that a species was observed at a given location but never that it was not observed at a given location. Consequently, a region without any observed specimen in the data remains highly uncertain. Some specimens could live there but were not observed, or no specimen live there but this information is not recorded. Finally, knowing abundance in space doesn't give information about the ecological determinants of species presence.

For all these reasons, SDM is usually achieved through *environmental niche modeling* approaches, i.e. by predicting the distribution in the geographic space on the basis of a representation in the environmental space. This environmental space is in most cases represented by climate data (such as temperature, and precipitation), but also by other variables such as soil type, land cover, distance to water, etc. Then, the objective is to learn a function that takes the environmental feature vector of a given location as input and outputs an estimate of the abundance of the species. The main underlying hypothesis is that the abundance function is related to the *fundamental ecological niche* of the species, in the sense of Hutchinson (see [1]). That means that in theory, a given species is likely to live in a single privileged ecological niche, characterized by an unimodal distribution in the environmental space. However, in reality, the abundance function is expected to be more complex. Many phenomena can actually affect the distribution of the species relative to its so called *abiotic* preferences. For instance, environment perturbations, or geographical constraints, or interactions with other living organisms (including humans) might have encourage specimens of that species to live in a different environment. As a consequence, the *realized ecological niche* of a species can be much more diverse and complex than its hypothetical fundamental niche.

## 10.1.2   Interest of Deep and Convolutional Neural Networks for SDM

**Notations**   When talking about environmental input data, there could be confusions between their different possible formats. Without precision given, $x$ will represent a general input environmental variable which can have any format. When a distinction is made, $x$ will represent a vector, while an array is always noted $X$. To avoid confusions on notations for the different index kinds, we note the spatial **site** index as superscript on the input variable ($x^k$ or $X^k$ for $k^{th}$ site) and the component index as subscript (so $x^k_j$ for the $j^{th}$ component of $k^{th}$ site vector $x_k \in \mathbb{R}^p$, or for the array $X^k \in \mathcal{M}_{d,e,p}(\mathbb{R})$, $X^k_{.,j,.}$ is the $j^{th}$ matrix slice taken on its second dimension). When we denote an input associated with a precise **point location** taken in a continuous spatial domain, the point $z$ is noted as argument: $x(z)$.

Classical SDM approaches postulate that the relationship between output and environmental variables is relatively simple, typically of the form:

$$g(\mathbb{E}[y|x]) = \sum_j f_j(x_j) + \sum_{j,j'} h_{j,j'}(x_j, x_{j'}) \tag{10.1}$$

where $y$ is the response variable targeted, a presence indicator or an abundance in our case, the $x_j$'s are components of a vector of environmental variables given as input for our model, $f_j$ are real monovariate functions of it, $h_{j,j'}$ are bivariate real functions representing pairwise interactions effects between inputs, and $g$ is a link function that makes sure $\mathbb{E}[y|x]$ lies in the space of our response variable $y$. State-of-the-art classification or regression models used for SDM in this way include GAM [2], MARS [3] or MAXENT [4, 5]. Thanks to $f_j$, we can isolate and understand the effect of the environmental factor $x_j$ on the response. Often, pairwise effects form of $h_{j,j'}$ is restricted to products, like it is the case in the very popular model MAXENT. It facilitates the interpretation and limits the dimensionality of model parameters. However, it sets a strong prior constraint without a clear theoretical founding as the explanatory factors of a species presence can be related to complex environmental patterns.

To overcome this limitation, deep feedforward neural networks (NN) [6] are good candidates, because their architecture favor high order interactions effects between the input variables, without constraining too much their functional form thanks to the depth of their architecture. To date, deep NN have shown very successful applications, in particular image classification [7]. Until now, to our knowledge, only one-layered-NN's have been tested in the context of SDM (e.g. in [8] or [9]). If they are able to capture a large panel of multivariate functions when they have a large number of neurons, their optimization is difficult, and deep NN have been shown empirically to improve optimization and performance (see section 6.4.1 in [6]). However, NN overfit seriously when dealing with small datasets, which is the case here ($\approx$5000 data), for this reason we need to find a way to regularize those models in a relevant way. An idea that is often used in SDM (see for example [10]) and beyond is to mutualize the heavy parametric part of the model for many species responses in order to reduce the space of parameters with highest likelihood. To put it another way, a NN that shares last hidden layer neurons for the responses of many species imposes a clear constraint: the parameters must construct high level ecological concepts which will explain as much as possible the abundance of all species. These high-level descriptors, whose number is controlled, should be seen as environmental variables that synthesize the most relevant information in the initial variables.

Another limitation of models described by Eq. (10.1) is that they don't capture spatial autocorrelation of species distribution, nor the information of spatial patterns described by environmental variables which can impact species presence. In the case of image recognition, where the explanatory data is an image, the variables, the pixels, are spatially correlated, as are the environmental variables used in the species distribution models. Moreover, the different channels of an image, RGB, can not be considered as being independent of the others because they are conditioned by the nature of the photographed object. We can see the environmental variables of a natural landscape in the same way as the channels of an image, noting that climatic, soil, topological or land use factors have strong correlations with others, they are basically not independent of each other. Some can be explained by common mechanisms as is the case with the different climatic variables, but some also act directly on others, as is the case for soil and climatic conditions on land use in agriculture, or the topology on the climate. These different descriptors can be linked by the concept of ecological environment. Thus, the heuristic that guides our approach is that the ecological niche of a species can be more effectively associated with high level ecological descriptors that combine non linearly the environmental variables on one hand, and the identification of multidimensional spatial patterns of images of environmental descriptors on the other hand. Convolutional neural networks (CNN, see [11]) applied to multi-dimensional spatial rasters of environmental variables can theoretically capture those, which makes them of particular interest.

### 10.1.3   Contribution

This work is the first attempt in applying deep feedforward neural networks and convolutional neural networks in particular to species distribution modeling. It introduces and evaluates several architectures based on a probabilistic modeling suited for regression on count data, the Poisson regression. Indeed, species occurrences are often spatially degraded in publicly available datasets so that it is statistically and computationally more relevant to aggregate them into counts. In particular, our experiments are based on the count data of the National Inventory for Nature Protection (INPN[1]), for 50 plant species over the metropolitan French territory along with various environmental data. Our models are compared to MAXENT, which is among the most used classical model in ecology. Our results first show how mutualizing model features for many species prevent deep NN to overfit and finally allow them to reach a better predictive performance than the MAXENT baseline. Then, our results show that convolutional neural networks performed even better than classical deep feedforward networks. This shows that spatially extended environmental patterns contain relevant extra information compared to their punctual values, and that species generally have a highly autocorrelated distribution in space. Overall, an important outcome of our study is to show that a restricted number of adequately transformed environmental variables can be used to predict the distribution of a huge number of species. We believe the study of the high-level environmental descriptors learned by the deep NNs could help to better understand the co-abundance of different species, and would be of great interest for ecologists.

## 10.2   A Deep Learning Model for SDM

### 10.2.1   A Large-Scale Poisson Count Model

In this part, we introduce the statistical model which we assume generates the observed data. Our data are species observations without sampling protocol and spatially aggregated on large spatial quadrat cells of $10 \times 10$ km. Thus, it is relevant to see them as counts.

To introduce our proposed model, we first need to clarify the distinction between the notion of "obsvered abundance" and "probability of presence". Abundance is a number of specimens relatively to an area. In this work, we model species *observed abundance* rather than *probability of presence* because we work with presence only data and without any information about the sampling process. Using presence-absence models, such as logistic regression, could be possible but it would

---

[1] http://https://inpn.mnhn.fr/.

require to arbitrarily generate absence data. And it has been shown that doing so can highly affect estimation and give biased estimates of total population [12]. Working with observed abundance doesn't bias the estimation as long as the space if homogeneously observed and we don't look for absolute abundance, but rather relative abundance in space.

The observed abundance, i.e. the number of specimens of a plant species found in a spatial area, is very often modeled by a Poisson distribution in ecology: when a large number of seeds are spread in the domain, each being independent and having the same probability of growing and being seen by someone, the number of observed specimens in the domain will behave very closely to a Poisson distribution. Furthermore, many recent SDM models, especially MAXENT as we will see later, are based on inhomogeneous Poisson point processes (IPP) to model the distribution of species specimens in an heterogeneous environment. However, when geolocated observations are aggregated in spatial quadrats ($\approx 10 \times 10$ km each in our case), observations must be interpreted as count per quadrats. If we consider $K$ quadrats named $(s_1, \ldots, s_K)$ (we will call them sites from now), with empty intersection, and we consider observed specimens are distributed according to $\mathscr{I}\mathscr{P}\mathscr{P}(\lambda)$, where $\lambda$ is a positive function defined on $\mathbb{R}^p$ and integrable over our study domain $D$ (where $x$ is known everywhere), we obtain the following equation:

$$\forall k \in [|1, K|], N(s_k) \sim \mathscr{P}\left(\int_{s_k} \lambda(x(z))dz\right) \qquad (10.2)$$

Now, in a parametric context, for the estimation of the parameters of $\lambda$, we need to evaluate the integral by computing a weighted sum of $\lambda$ values taken at quadrature points representing all the potential variation of $\lambda$. As our variables $x$ are constant by spatial patches, we need to compute $\lambda$ on every point with a unique value of $x$ inside $s_k$, and to do this for every $k \in [|1, K|]$. This can be very computationally and memory expensive. For example, if we take a point per square km (common resolution for environmental variables), it would represent 518,100 points of vector, or patch, input to extract from environmental data and to handle in the learning process. At the same time, environmental variables are very autocorrelated in space, so the gain in estimation quality can be small compared to taking a single point per site. Thus, for simplicity, we preferred to make the assumption, albeit coarse, that the environmental variables are constant on each site and we take the central point to represent it. Under this assumption, we justify by the following property the Poisson regression for estimating the intensity of an IPP.

**Property** The inhomogeneous Poisson process estimate is equivalent to a Poisson regression estimate with the hypothesis that $x(z)$ is constant in any given site of the domain.

**Proof** We note $z_1, \ldots, z_N \in D$ the $N$ species observations points, $K$ the number of disjoints sites making a partition of $D$, and assumed to have an equal area. We write the likelihood of $z_1, \ldots, z_N$ according to the inhomogeneous poisson process of intensity function $\lambda \in (\mathbb{R}^+)^D$:

$$p(z_1, \ldots, z_N | \lambda) = p(N | \lambda) \prod_{i=1}^{N} p(z_i | \lambda)$$

$$= \frac{(\int_D \lambda)^N}{N!} \exp\left(-\int_D \lambda\right) \prod_{i=1}^{N} \frac{\lambda(x(z_i))}{\int_D \lambda}$$

$$= \frac{\exp\left(-\int_D \lambda\right)}{N!} \prod_{i=1}^{N} \lambda(x(z_i))$$

We transform the likelihood with the logarithm for calculations commodity:

$$\log(p(z_1, \ldots, z_N | \lambda)) = \sum_{i=1}^{N} \log\left(\lambda(x(z_i))\right) - \int_D \lambda - \log(N!)$$

We leave the $N!$ term, as it has no impact on the optimisation of the likelihood with respect to the parameters of $\lambda$. Now, $\int_D \lambda$ simplifies to a sum, as $x(z)$ is constant inside each site of $D$ :

$$\sum_{i=1}^{N} \log\left(\lambda(x(z_i))\right) - \int_D \lambda = \sum_{i=1}^{N} \log\left(\lambda(x(z_i))\right) - \sum_{k \in \text{Sites}} \frac{|D|}{K} \lambda(x^k)$$

$$= \sum_{k \in \text{Sites}} n_k \log\left(\lambda(x^k)\right) - \frac{|D|}{K} \lambda(x^k)$$

Where $n_k$ is the number of species occurrences that fall in site $k$. We can aggregate the occurrences that are in a same site because $x$ is the same for them. We can now factorize $|D|/K$ on the whole sum, which brings us, up to the factor, to the poisson regression likelihood with pseudo-counts $K n_k / |D|$.

$$= \frac{|D|}{D} \sum_{k \in \text{Sites}} \frac{D n_k}{|D|} \log\left(\lambda(x^k)\right) - \lambda(x^k)$$

So maximizing this log-likelihood is exactly equivalent to maximizing the initial Poisson process likelihood.                                                                    $\square$

Proof uses the re-expression of the IPP likelihood, inspired from [13], as that of the associated Poisson regression. In the following parts, we always consider that, for a given species, the number $y$ of specimens observed in a site of environmental input $x$ is as follows:

$$y \sim \mathscr{P}(\lambda_{m,\theta}(x))  \tag{10.3}$$

Where $m$ is a model architecture with parameters $\theta$.

From Eq. (10.3), we can write the likelihood of counts on $K$ different sites $(x_1, \ldots, x_K)$ for $N$ independently distributed species with abundance functions $\lambda_{m_1,\theta_1}, \ldots, \lambda_{m_N,\theta_N} \in (\mathbb{R}^+)^{\mathbb{R}^p}$, respectively determined by models $(m_i)_{i\in[|1,N|]}$ and parameters $(\theta_i)_{i\in[|1,N|]}$:

$$p\left((y_k^i)_{i\in[|1,N|],k\in[|1,K|]} | (\lambda_{m_i,\theta_i})_{i\in[|1,N|]}\right) = \prod_{i=1}^{N}\prod_{k=1}^{K} \frac{(\lambda_{m_i,\theta_i}(x_k))^{y_k^i}}{y_k^i!} \exp(-\lambda_{m_i,\theta_i}(x_k))$$

Which gives, when eliminating $\log(y_k^i)!$ terms (which are constant relatively to models parameters), the following negative log-likelihood :

$$\mathscr{L}\left((y_k^i)_{i\in[|1,N|],k\in[|1,K|]} | (\lambda_{m_i,\theta_i})_{i\in[|1,N|]}\right) := \sum_{i=1}^{N}\sum_{k=1}^{K} \lambda_{m_i,\theta_i}(x_k) - y_k^i \log(\lambda_{m_i,\theta_i}(x_k))$$

$$\tag{10.4}$$

Following the principle of maximum likelihood, for fitting a model architecture, we minimize the objective function given in Eq. (10.4) relatively to parameters $\theta$.

## 10.2.2 Links with MAXENT

For our experiment, we want to compare our proposed models to a state of the art method commonly used in ecology. We explain in the following why and how we can compare the chosen reference, MAXENT, with our models.

MAXENT [4, 5] is a popular SDM method and related software for estimating relative abundance as a function of environmental variables from presence only data points. This method has proved to be one of the most efficient in prediction [14], while guaranteeing a good interpretability thanks to the simple elementary form of its features and its variable selection procedure. The form of the relative abundance function belongs to the class described in Eq. (10.1). More specifically:

$$\log\left(\lambda_{MAX,\theta}(x)\right) = \alpha + \sum_{j=1}^{p}\sum_{s=1}^{S} f_j^s(x_{(j)}) + \sum_{j<j'} \beta_{j,j'} x_j x_j'  \tag{10.5}$$

where $x_{(j)}$ is the $j^{th}$ component of vector $x$. The link function is a logarithm, and variables interactions effects are product interactions. If $x_j$ is a quantitative variable the functions $(f_s)_{s \in [|1,S|]}$ belongs to four categories: linear, quadratic, threshold and hinge. One can get details on the hinges functions used in MAXENT in [15]. If $x_j$ is categorical, then $f_j$ takes a different value for every category, with one zero category.

It has been shown that MAXENT method is equivalent to the estimation of an IPP intensity function with a specific form and a weighted L1 penalty on its variables [16]. Let's call $\lambda_{MAX,\theta}(x)$ the intensity predicted by MAXENT with parameters $\theta$ at $x$. Last property says that on any given dataset, $\hat{\theta}$ estimated from a Poisson regression (aggregating observations as counts per site) is the same as the one of the IPP (each observation is an individual point, even when there are several at a same site). In our experiments, we ran MAXENT using the `maxnet` package in R [17], with the default regularization, and giving to the function :

1. A positive point per observation of the species.
2. A pseudo-absence point per site.

MAXENT returns only the parameters of the $(f_j^s)_{s,j}$ and the $(\beta_{j,j'})_{j<j'}$, but not the intercept $\alpha$, as it is meant to only estimate the absolute abundance. We don't aim at estimating absolute abundance either, however, we need the intercept to measure interesting performance metrics across all the compared models. To resolve this, for each species, we fitted the following model using the `glm` package in R as a second step:

$$y \sim \mathscr{P}\left(\exp(\alpha + \log(p))\right)$$

Where $\alpha$ is our targeted intercept, $p$ is the relative intensity prediction given by MAXENT at the given site, and $y$ is the observed number of specimens at this site.

## 10.2.3   SDM Based on a Fully-Connected NN Model

We give in the following a brief description of the general structure of fully-connected NN models, and how we decline it in our tested deep model architecture.

### 10.2.3.1   General Introduction of Fully-Connected NN Models

A deep NN is a multi-layered model able to learn complex non-linear relationship between an input data, which in our case will be a vector $x \in \mathbb{R}^p$ of environmental variables that is assumed to represent a spatial site, and output variables $y_1, \ldots, y_N$, which in our case is species counts in the spatial site. The classic so called **fully-connected** NN model is composed of one or more **hidden layer(s)**, and each layer is composed of one or more **neuron(s)**. We note $n(l, m)$ the number of neurons of layer $l$ in model architecture $m$. $m$ parameters are stored in $\theta$. In the first layer, each neuron is the result of a parametric linear combination of the elements of $x$, which is then transformed by an **activation function** $a$. So for a NN $m$, $a_m^{1,j}(x, \theta) := a(x^T \theta_j^1)$ is called **the activation** of $j^{th}$ neuron of the first hidden layer of $m$ when it is applied to $x$. Thus, on the $l^{th}$ layer with $l > 1$, the activation of the $j^{th}$ neuron is $a((\theta_j^l)^T a_m^{l-1,\cdot\cdot})$. Now, we understand that the neuron is the unit that potentially combines every variables in $x$, and, its activation inducing a non-linearity to the parametric combination, it can be understood as a particular basis function in the $p$ dimensional space of $x$. Thus, the model is able to combine as many basis functions as there are neurons in each layer, and the basis functions become more and more complex when going to further layers. Finally, these operations makes $m$ theoretically able to closely fit a broad range of functions of $x$.

Learning of model parameters is done through optimization (minimization by convention) of an objective function that depends on the prediction goal. Optimization method for NN parameters $\theta$ is based on stochastic gradient descent algorithms, however, the loss function gradient is approximated by the back-propagation algorithm [18].

Learning a NN model lead to a lot of technical difficulties that have been progressively dealt with during last decade, and through many different techniques. We present some that have been of particular interest in our study. A first point is that there are several types of activation functions, the first one introduced being the sigmoid function. However, the extinction of its gradient when $x^T \theta_j^1$ is small or big, has presented a serious problem for parameters optimization in the past. More recently, the introduction of the ReLU [19] activation function helped made an important step forward in NNs optimization. A second point is that when we train a NN model, simultaneous changes of all the parameters lead to important change in the distribution (across the dataset) of each activation of the model. This phenomenon is called internal covariate shift, and perturbs learning importantly. Batch-Normalization [20] is a technique that significantly reduces internal covariate shift and help to regularize our model as well. It consists of a parameterized centering and reduction of pre-activations. This facilitates optimization and enables to raise the learning rate leading to a quicker convergence. At the same time, it has a regularization effect because the centering and reduction of a neuron activation is linked to the mini-batch statistics. The mini-batch selection being stochastic at every iteration, a neuron activation is stochastic itself, and the model will not rely on it when it has no good effect on prediction.

### 10.2.3.2 Models Architecture in This Study

For a given species $i$, When we know the model parameter $\theta$, we can predict the parameter of the Poisson distribution of the random response variable $y_i \in \mathbb{N}$, i.e. the count of species $i$, conditionally on its corresponding input $x$, with the formula :

$$\lambda_{m,\theta}(x) = \exp(\gamma_i^T a_m^{N_h,\cdot\cdot}(x, \theta)) \tag{10.6}$$

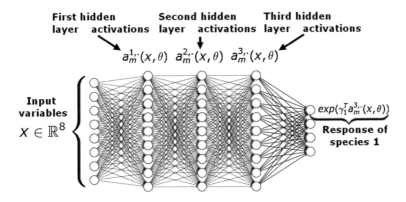

**Fig. 10.1** A schematic representation of fully-connected NN architecture. Except writings, image comes from Michael®Nielsen[2].

For this work, we chose the logarithm as link function $g$ mentioned in **1.2**. It is the conventional link function for the generalized linear model with Poisson family law, and is coherent with MAXENT. $\gamma_i \in \mathbb{R}^{n(N_h, m)}$ is included in $\theta$. It does the linear combinations of last layer neurons activations for the specific response $i$. If we set $n(N_h, m) := 200$ as we do in the following experiments, there are only 200 parameters to learn per individual species, while there are a lot more in the shared part of the model that builds $a_m^{N_h \cdot \cdot}(x, \theta)$. Now for model fitting, we follow the method of the maximum likelihood, **the objective function will be a negative-loglikelihood**, but it could otherwise be some other prediction error function. Note that we will rather use the term **loss function** than negative loglikelihood for simplicity. We chose **the ReLU as activation function**, because it showed empirically less optimization problems and a quicker convergence. Plus, we empirically noticed the gain in optimization speed and less complications with the learning rate initialization when using Batch-Normalization. For this reason, Batch-Normalization is applied to every pre-activation (before applying the ReLU) to every class of NN model in this paper, even with CNNs. We give a general representation of the class of NN models used in this work in Fig. 10.1.

### 10.2.4   SDM Based on a Convolutional NN Model

A convolutional NN (CNN) can be seen as a extension of NN that are particularly suited to deal with certain kind of input data with very large dimensions. They are of particular interest in modeling species distribution, because they are able to capture the effect of spatial environmental patterns. Again, we will firstly describe the general form of CNN before going to our modeling choices.

---

[2]http://neuralnetworksanddeeplearning.com/chap6.html

### 10.2.4.1 General Introduction of CNN Models

CNN is a form of neural network introduced in [11]. It aims to efficiently apply NN to input data of large size (typically 2D or 3D arrays, like images) where elements are spatially auto-correlated. For example, using a fully-connected neural network with 200 neurons on an input RGB image of dimensions $256 \times 256 \times 3$ would imply around $4 * 10^7$ parameters only for the first layer, which is already too heavy computationally to optimize on a standard computer these days. Rather than applying a weight to every pixel of an input array, CNN will apply a **parametric discrete convolution**, based on a kernel of reasonable size (3/3/p or 5/5/p are common for N/N/p input arrays) on the input arrays to get an intermediate feature map (2D). The convolution is applied with a moving windows as illustrated in Fig. 10.2b. Noting $\mathbf{X} \in \mathcal{M}_{d,d,p}$ an input array, we simplify notations in all that follows by writing $\mathscr{CV}(X, k_\gamma(c))$ the resulting feature map from applying the convolution with $(c, c, p)$ kernel of parameters $\gamma \in \mathbb{R}^{c^2 p}$. If the convolution is applied directly on $\mathbf{X}$, the sliding window will pass its center over every $X_{i,j,.}$ from the up-left to the bottom-right corner and produce a feature map with a smaller size than the input because $c > 1$. The **zero-padding operation** removes this effect by adding $(c - 1)/2$ layers of 0 on every side of the array. After a convolution, there can be a Batch-Normalization and an activation function is generally applied to each pixel of the features maps. Then, there is a synthesizing step made by the **pooling** operation. Pooling aggregates groups of cells in a feature map in order to reduce its size and introduce invariance to local translations and distortions. After having composed these operations several times, when the size of feature maps is reasonably small (typically reaching 1 pixel), a **flattening** operation is applied to transform the 3D array containing all the feature maps into a vector. This features vector will then be given as input to a fully-connected layer as we described in last part. The global concept underlying convolution layers operations is that first layers act as low level interpretations of the signal, leading to activations for salient or textural patterns. Last layers, on their side, are able to detect more complex patterns, like eyes or ears in the case of a face picture. Those high levels features have much greater sense regarding predictions we want to make. Plus, they are of much smaller dimension than the input data, which is more manageable for a fully-connected layer.

### 10.2.4.2 Constitution of a CNN Model for SDM

The idea which pushes the use of CNN models for SDM is that complex spatial patterns like a water network, a valley, etc., can affect importantly the species abundance. This kind of pattern can't be really deducted for punctual values of environmental variables. Thus, we have chosen to build a SDM model which takes as input an array with a map of values for each environmental variable that is used in the other models. This way, we will be able to conclude if there is extra relevant information in environmental variables spatial patterns to predict better species distribution. In Fig. 10.2a, we show for a single site a subsample of environmental variables maps taken as input by our CNN model. To provide some more detail about the model architecture, the input array $X$ is systematically padded such that the feature map resulting from the convolution is of same size as 2 first dimensions of the input ($(c - 1)/2$ cells of 0 after on the sides of the 2 dimensions). To illustrate that, our padding policy is the same as the one illustrated in the example given in Fig. 10.2b. However, notice that the kernel size can differ and the third dimension size of input array will be the number of input variables or feature maps. For an example of For the reasons described in **2.3, we applied a Batch-Normalization** to each feature map (same normalization for every pixels of a map) before the activation, which **is still a ReLU**. For the pooling operation, we chose the **average pooling** which seems intuitively more relevant to evaluate an abundance (=concentration). The different kinds of operations and their succession in our CNN model are illustrated in Fig. 10.2c.

## 10.3   Data and Methods

### *10.3.1   Observations Data of INPN*

This paper is based on a reference dataset composed of count data collected and validated by French expert naturalists. This dataset, referred as INPN[3] for "national inventory of natural heritage" [21], comes from the GBIF portal.[4] It provides access to occurrences data collected in various contexts including Flora and regional catalogs, specific inventories, field note books, and prospections carried out by the botanical conservatories. In total, the INPN data available on the GBIF contains 20,999,334 occurrences, covering 7626 species from which we selected 1000 species.

The assets of this data are the quality of their taxonomic identification (provided by an expert network), their volume and geographic coverage. Its main limitation, however, is that the geolocation of the occurrences was degraded (for plant protection concerns). More precisely, all geolocations were aggregated to the closest central point of a spatial grid composed of $100\,km^2$ quadrat cells (i.e. sites of $10\times10\,km$). Thus, the number of observations of a species falling in a site gives a count.

In total, our study is based on 5181 sites, which are split in 4781 training sites for fitting models, and 400 test sites for validating and comparing models predictions.

---

[3]https://inpn.mnhn.fr.

[4]https://www.gbif.org/.

**a.** Example of input environmental array.

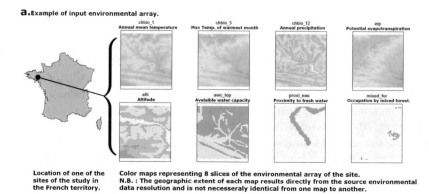

Location of one of the sites of the study in the French territory.

Color maps representing 8 slices of the environmental array of the site.
N.B. : The geographic extent of each map results directly from the source environmental data resolution and is not necesseraly identical from one map to another.

**b.** Operations specific to CNN: Convolution, pooling and flattening.

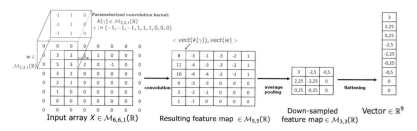

**C.** Schematic structure of convolutional layers.

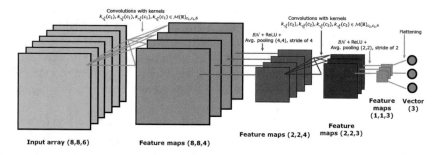

**Fig. 10.2** (**a**) Examples of input environmental data (**b**) for convolution, pooling and flattening process in our (**c**) Convolutional Neural Network architecture

## 10.3.2 Species Selection

For the genericity of our results and to make sure they are not biased by the choice of a particular category of species, we have chosen to work with a high number of randomly chosen species. From the 7626 initial species, we selected species with more than 300 observations. We selected amongst those a random subset of 1000 species to constitute an ensemble $E_{1000}$. Then, we randomly selected 200 species amongst $E_{1000}$ to constitute $E_{200}$, and finally randomly selected 50 in $E_{200}$ which gave $E_{50}$. $E_{50}$ being the main dataset used to compare our model to the baselines, we provide in Fig. 10.1 the list of species composing it. The full dataset with species of $E_{1000}$ contains 6,134,016 observations in total (see Table 10.1 for the detailed informations per species).

### 10.3.3 Environnemental Data

In the following, we denote by $p$ the number of environmental descriptors. For this study, we gathered and compiled different sources of environmental data into $p = 46$ geographic rasters containing the pixel values of environmental descriptors presented in Table 10.2 with several resolutions, nature of values, but having a common cover all over the metropolitan French territory. We chose some typical environmental descriptors for modeling plant distribution that we believe carry relevant information both as punctual and spatial representation. They can be classified as bioclimatic, topological, pedologic hydrographic and land cover descriptors. In the following, we briefly describe the sources, production method, and resolution of initial data, and the contingent specific post-process for reproducibility.

#### 10.3.3.1 Climatic Descriptors: Chelsea Climate Data 1.1

Those are raster data with worldwide coverage and 1 km resolution. A mechanistical climatic model is used to make spatial predictions of monthly mean-max-min temperatures, mean precipitations and 19 bioclimatic variables, which are downscaled with statistical models integrating historical measures of meteorologic stations from 1979 to today. The exact method is explained in the reference papers [22] and [23]. The data is under Creative Commons Attribution 4.0 International License and downloadable at (http://chelsa-climate.org/downloads/).

#### 10.3.3.2 Potential Evapotranspiration: CGIAR-CSI ETP Data

The CGIAR-CSI distributes this worldwide monthly potential-evapotranspiration raster data. It is pulled from a model developed by Antonio Trabucco [24, 25]. Those are estimated by the Hargreaves formula, using mean monthly surface temperatures and standard deviation from WorldClim 1:4 (http://www.worldclim. org/), and radiation on top of atmosphere. The raster is at a 1km resolution, and is freely downloadable for a nonprofit use at:

http://www.cgiar-csi.org/data/global-aridity-and-pet-database#description

**Table 10.1** List of species in $E_{50}$ with the total number of observations and prevalence in the full database

| Taxon name | Total # obs. | Prevalence |
|---|---|---|
| *Alisma plantago-aquatica* L. | 15,324 | 56.3 |
| *Alopecurus geniculatus* L. | 5703 | 31.5 |
| *Antennaria carpatica* (Wahlenb.) Bluff & Fingerh | 1780 | 4.0 |
| *Anthrisen sylvestris* (L.) Hoffm. | 27,381 | 64.9 |
| *Astragalus hypoglottis* L. | 1901 | 5.7 |
| *Berteroa incana* (L.) DC. | 3966 | 11.2 |
| *Biscutella brevicaulis* Jord. | 450 | 1.0 |
| *Campanula spicata* L. | 544 | 1.7 |
| *Carduus vivariensis* Jord. | 1577 | 7.4 |
| *Carex ericctorum* Pollich | 538 | 1.8 |
| *Carlina acanthifolia* All. | 6214 | 10.6 |
| *Centranthus augustifolius* (Mill.) DC. | 2755 | 5.9 |
| *Cladanthus mixtus* (L.) Chevall. | 637 | 5.3 |
| *Coronilla coronata* L. | 325 | 0.9 |
| *Cynoglossum creticum* Mill. | 1470 | 9.2 |
| *Cytisus villosus* Pourr. | 562 | 1.0 |
| *Dianthus pyrenaicus* Pourr. | 392 | 0.8 |
| *Epilobium alpestre* (Jacq.) Krocker | 1197 | 3.5 |
| *Euphorbia dendroide* L. | 747 | 0.5 |
| *Festuca cinerea* Vill. | 3795 | 5.3 |
| *Galium lucidum* All. | 3204 | 11.7 |
| *Galium timeroyi* Jord. | 1362 | 6.6 |
| *Helictotrichon sedenense* (Clarion ex DC.) Holub | 8498 | 5.4 |
| *Hieracium lawsonii* Vill. | 629 | 3.2 |
| *Hieracium praecox* Sch.Bip. | 998 | 4.7 |
| *Iris lutescens* Lam. | 2537 | 6.6 |
| *Juncus trifidus* L. | 3570 | 3.9 |
| *Lathyrus niger* (L.) Bernh. | 2474 | 13.8 |
| *Myrtus communis* L. | 2054 | 1.9 |
| *Meconopsis cambrica* (L.) Vig. | 1291 | 3.8 |
| *Oxalis corniculata* L. | 5628 | 37.5 |
| *Oxytropis fetida* (Vill.) DC. | 315 | 1.0 |
| *Persicaria vivipara* (L.) Rouse Decraene | 11,122 | 5.9 |
| *Phleum alpinurn* L. | 7267 | 6.3 |
| *Potamogeton coloratus* Hornem. | 813 | 5.5 |
| *Potentilla pusilla* Host | 655 | 1.7 |
| *Primula latifolia* Lapeyr. | 1268 | 1.8 |
| *Psilurus incurvus* (Gouan) Schinz & Thell. | 597 | 4.2 |
| *Ranunculus parnassifolius* L. | 371 | 1.0 |
| *Ranunculus repens* L. | 76,346 | 83.0 |
| *Reseda lutea* L. | 16,756 | 49.0 |

(continued)

**Table 10.1**  (continued)

| Taxon name | Total # obs. | Prevalence |
|---|---|---|
| *Rorippa pyrenaica* (All.) Rchb. | 2169 | 9.2 |
| *Rubus ulmifolius* Schott | 14,523 | 35.5 |
| *Thalictrum aquilegifolium* L. | 2855 | 8.8 |
| *Thalictrum alpinum* L. | 581 | 1.0 |
| *Trifolium micranthum* Viv. | 767 | 8.0 |
| *Valerianella rimosa* Bast. | 1518 | 13.8 |
| *Vicia onobrychioides* L. | 1602 | 6.3 |
| *Viola lactea* Sm. | 520 | 4.7 |
| *Visearia vulgaris* Bernh. | 781 | 3.2 |

### 10.3.3.3  Pedologic Descriptors: The ESDB v2: 1 km × 1 km Raster Library

The library contains multiple soil pedology descriptor raster layers covering Eurasia at a resolution of 1 km. We selected 11 descriptors from the library. More precisely, those variables have ordinal format, representing physico-chemical properties of the soil, and come from the PTRDB. The PTRDB variables have been directly derived from the initial soil classification of the Soil Geographical Data Base of Europe (SGDBE) using expert rules. SGDBE was a spatial relational data base relating spatial units to a diverse pedological attributes of categorical nature, which is not useful for our purpose. For more details, see [26, 27] and [28]. The data is maintained and distributed freely for scientific use by the European Soil Data Centre (ESDAC) at http://eusoils.jrc.ec.europa.eu/content/european-soil-database-v2-raster.

### 10.3.3.4  Altitude: USGS Digital Elevation Data

The Shuttle Radar Topography Mission achieved in 2010 by Endeavour shuttle managed to measure digital elevation at three arc second resolution over most of the earth surface. Raw measures have been post-processed by NASA and NGA in order to correct detection anomalies. The data is available from the U.S. Geological Survey, and downloadable on the Earthexplorer (https://earthexplorer.usgs.gov/). One can refer to https://lta.cr.usgs.gov/SRTMVF for more informations.

### 10.3.3.5  Hydrographic Descriptor: BD Carthage v3

BD Carthage is a spatial relational database holding many informations on the structure and nature of the french metropolitan hydrological network. For the purpose of plants ecological niche, we focus on the geometric segments representing

**Table 10.2** Table of 46 environmental variables used in this study

| Name | Description | Nature | Values | Resolution |
|------|-------------|--------|--------|------------|
| CHBIO_1 | Annual mean temperature | quanti. | [−10.6, 18.4] | 30 |
| CHBIO_2 | Mean of monthly max(temp)-min(temp) | quanti. | [7.8,21.0] | 30 |
| CHBIO_3 | Isothermality (100*chbio_2/chbio_7) | quanti. | [41.2,60.0] | 30 |
| CHBIO_4 | Temperature seasonality (std. dev.*100) | quanti. | [302,778] | 30 |
| CHBIO_5 | Max temperature of warmest month | quanti. | [36.4,6.2] | 30 |
| CHBIO_6 | Min temperature of coldest month | quanti. | [−28.2, 5.3] | 30 |
| CHBIO_7 | Temperature annual range (5–6) | quanti. | [16.7,42.0] | 30 |
| CHBIO_8 | Mean temperature of wettest quarter | quanti. | [−14.2, 23.0] | 30 |
| CHBIO_9 | Mean temperature of driest quarter | quanti. | [−17.7, 26.5] | 30 |
| CHBIO_10 | Mean temperature of warmest quarter | quanti. | [−2.8, 26.5] | 30 |
| CHBIO_11 | Mean temperature of coldest quarter | quanti. | [−17.7, 11.8] | 30 |
| CHBIO_12 | Annual precipitation | quanti. | [318,2543] | 30 |
| CHBIO_13 | Precipitation of wettest month | quanti. | [43.0,285.5] | 30 |
| CHBIO_14 | Precipitation of driest month | quanti. | [3.0,135.6] | 30 |
| CHBIO_15 | Precipitation seasonality (Coef. of Var.) | quanti. | [8.2,26.5] | 30 |
| CHBIO_16 | Precipitation of wettest quarter | quanti. | [121,855] | 30 |
| CHBIO_17 | Precipitation of driest quarter | quanti. | [20,421] | 30 |
| CHBIO_18 | Precipitation of warmest quarter | quanti. | [19.8,851.7] | 30 |
| CHBIO_19 | Precipitation of coldest quarter | quanti. | [60.5,520.4] | 30 |
| etp | Potential evapotranspiration transpiration | quanti. | [133,1176] | 30 |
| alti | Elevation | quanti. | [−188, 4672] | 3 |
| awc_top | Topsoil available water capacity | ordinal | {0, 120, 165, 210} | 30 |
| bs_top | Base saturation of the topsoil | ordinal | {35, 62, 85} | 30 |
| cec_top | Topsoil cation exchange capacity | ordinal | {7, 22, 50} | 30 |
| crusting | Soil crusting class | ordinal | [\|0, 5\|] | |
| dgh | Depth to a gleyed horizon | ordinal | {20, 60, 140} | 30 |
| dimp | Depth to an impermeable layer | ordinal | {60, 100} | 30 |
| erodi | Soil erodibility class | ordinal | [\|0, 5\|] | 30 |
| oc_top | Topsoil organic carbon content | ordinal | {1, 2, 4, 8} | 30 |
| pd_top | Topsoil packing density | ordinal | {1, 2} | 30 |
| text | Dominant surface textural class | ordinal | [\|0,5\|] | 30 |
| proxi_eau | <50 meters to fresh water | bool. | {0, 1} | 30 |
| arti | Artificial area: clc ∈ {1, 10} | bool. | {0, 1} | 30 |
| semi_arti | Semi-artificial area: clc ∈ {2, 3, 4, 6} | bool. | {0, 1} | 30 |
| arable | Arable land: clc ∈ {21, 22} | bool. | {0, 1} | 30 |
| pasture | Pasture land: clc ∈ {18} | bool. | {0, 1} | 30 |
| brl_for | Broad-leaved forest: clc ∈ {23} | bool. | {0, 1} | 30 |
| coni_for | Coniferous forest: clc ∈ {24} | bool. | {0, 1} | 30 |
| mixed_for | Mixed forest: clc ∈ {25} | bool. | {0, 1} | 30 |
| nat_grass | Natural grasslands: clc ∈ {26} | bool. | {0, 1} | 30 |
| moors | Moors: clc ∈ {27} | bool. | {0, 1} | 30 |

(continued)

**Table 10.2** (continued)

| Name | Description | Nature | Values | Resolution |
|------|-------------|--------|--------|------------|
| sclero | Sclerophyllous vegetation: clc $\in$ {28} | bool. | {0, 1} | 30 |
| transi_wood | Transitional woodland-shrub: clc $\in$ {29} | bool. | {0, 1} | 30 |
| no_veg | No or few vegetation: clc $\in$ {31, 32} | bool. | {0, 1} | 30 |
| coastal_area | Coastal area: clc $\in$ {37, 38, 39, 42, 30} | bool. | {0, 1} | 30 |
| ocean | Ocean surface: clc $\in$ {44} | bool. | {0, 1} | 30 |

watercourses, and polygons representing hydrographic fresh surfaces. The data has been produced by the *Institut National de l'information Géographique et forestière* (IGN) from an interpretation of the BD Ortho IGN. It is maintained by the SANDRE under free license for non-profit use and downloadable at:
http://services.sandre.eaufrance.fr/telechargement/geo/ETH/BDCarthage/FX
From this shapefile, we derived a raster containing the binary value of variable proxi_eau, i.e. proximity to fresh water, all over France. We used qgis to rasterize to a 12.5 m resolution, with a buffer of 50 m, the shapefile COURS_D_EAU.shp on one hand, and the polygons of
*SURFACES_HYDROGRAPHIQUES.shp with attribute NATURE="Eau douce permanente" on the other hand. We then created the maximum raster of the previous ones (So the value of 1 correspond to an approximate distance of less than 50 m to a watercourse or hydrographic surface of fresh water).

### 10.3.3.6   Land Cover: Corine Land Cover 2012, Version 18.5.1, 12/2016

It is a raster layer describing soil occupation with 48 categories across Europe (25 countries) at a resolution of 100 m. This classification is the result of an interpretation process from earth surface high resolution satellite images. This data base of the European Union is freely accessible online for all use at http://land.copernicus.eu/pan-european/corine-land-cover/clc-2012 and commonly used for the purpose of plant distribution modeling. For a need of meaningfull variables at our scale and reduced memory consumption, we reduced the number of categories to 14 following mainly the procedure of They eliminate some categories of few interest, too rare or inaccurate, and groups categories that are associated with similar plant communities. In addition, we introduce a category "Semi artificial surfaces", which regroups perturbed natural areas, interesting for the study of alien invasive species. We keep the category "Sea and ocean" from the Corine Land Cover classification because it can be an important contextual variable for the convolutional neural network model. The final categories groups are detailed in Table 10.2. for each of the retain categories, we created a raster of the same resolution as the original one, where the value 1 means the pixel belongs to the category, or the value is 0 otherwise.

#### 10.3.3.7 Environmental Variables Extraction and Format

When creating the $p$ global GeoTIIF rasters, as the original coordinate system of the layer vary among sources, we change it if necessary to WGS84 using `rgdal` package on R, which is the coordinate system INPN occurrences databases. As explained previously, for computational reasons considering the scale, and simplicity, we chose to represent each site by a single geographic point, and chose the center of the site. We are going to compare two types of models. For a site $k$, the first takes as input a vector of $p$ elements which values are those of the environmental variables taken at the geolocation of the center of the site $k$, while the other takes $p$ rasters of size (d,d) cropped (with package `raster`) from the global raster of each environmental descriptors and centered at the center of $k$. If we denote $res_{\text{lon},j}$ the spatial resolution in longitude of global raster of the $j_t h$ environmental descriptor, and $res_{\text{lat},j}$ its resolution in latitude, the spatial extent of $X_{.,.,j}^k$ is $(d.res_{\text{lat},j} \times d.res_{\text{lon},j})$. As a consequence, the extents are heterogeneous across environmental descriptors. In this study, we experimented the method with $d = 64$, so the input data items $X^k$ learned by our convolutional model is of dimension $64 \times 64 \times 46$.

### 10.3.4 Detailed Models Architectures and Learning Protocol

MAXENT is learned independently on every species of $E_{50}$. Similarly, we fit a classic loglinear model to give a naive reference. Then, two architectures of NN are tested, one with a single hidden layer (SNN), one with six hidden layers (DNN). Those models take a vector of environmental variables $x^k$ as input. As introduced previously, we want to evaluate if training a multi-response NN model, i.e. a NN predicting several species from a single $a_m^{N_h(m)}(x, \theta)$, can prevent overfitting. One architecture of CNN is tested, which takes as input an array $X^k$. Hereafter, we described more precisely the architecture of those models.

#### 10.3.4.1 Baseline Models

- **LGL** Considering a site $k$, and its environmental variables vector $x^k$, the output function $\lambda_{LGL}$ of the loglinear model parametrized by $\beta \in \mathbb{R}^p$ is simply the exponential of a scalar product between $x^k$ and $\beta$ :

$$\lambda_{LGL}(x^k, \beta) = \exp\left(\beta^T x^k\right)$$

As LGL has no hidden layer, we learned a multi-response model, which is equivalent to fitting the 50 mono-response models independently.

- **MAXENT**.

#### 10.3.4.2  Proposed Models Based on NN

- **SNN** has only 1 hidden layer ($N_h = 1$) with 200 neurons ($|a_{SNN}^1| = 200$) all batch-normalized and the activation function is ReLU. As the architecture is not deep, it makes a control example to evaluate when stacking more layers. SNN is tested in 3 multi-response versions, on $E_{50}$, $E_{200}$ or $E_{1000}$.
- **DNN** is a deep feedforward network with $N_h = 6$ hidden layers and $n(l, DNN) = 200, \forall l \in [|1, 6|]$. Every pre-activation is Batch-normalized and has a ReLU activation. DNN is tested in 4 versions, the mono-response case fitted independently on each species of $E_{50}$ like MAXENT and LGL, and the multi-response fitted on $E_{50}$, $E_{200}$ or $E_{1000}$.
- **CNN** is composed of two hidden convolutional layers and one last layer fully connected with 200 neurons, exactly similar to previous ones. The first layer is composed of 64 convolution filters of kernel size $(3, 3)$ and 1 line of 0 padding. The resulting feature maps are batch-normalized (same normalization for every pixels of a feature map) and transformed with a Relu. Then, an average pooling with a $(8, 8)$ kernel and $(8, 8)$ stride is applied. The second layer is composed of 128 convolution filters of kernel size $(5, 5)$ and 2 lines of padding, plus Batch-Normalization and ReLU. After, that a second average pooling with a $(8, 8)$ kernel and $(8, 8)$ kernel and $(8, 8)$ stride reduces size of the 128 feature maps to one pixel. Those are collected in a vector by a flattening operation preceding the fully connected layer. This architecture is not very deep. However, considered the restricted number of samples, a deep CNN would be very prone to over fitting. CNN is tested in multi-responses versions on $E_{50}$, $E_{200}$ and $E_{1000}$.

#### 10.3.4.3  Models Optimization

Our experiments were conducted using the R framework (version 3.3.2), on a Windows 10 machine with 2 CPUs with 2.60 GHz and 4 cores each, and one GPU NVIDIA Quadro M1000M. mxnet [29] is a convenient C++ library for learning deep NN models and is deployed as an R package. It integrates a high level symbolic language for quickly building customized models and loss functions, and automatically distributes calculations under CPUs or GPUs.

We fit the MAXENT model for every species of $E_{50}$ with the recently released R package maxnet [17] and the vector input variables.

The LGL model was fitted with the package mxnet. The loss being convex, we used a simple **gradient descent algorithm** and stopped when the gradient norm was close to 0. The learning took around 2 min.

SNN, DNN and CNN models are fitted with the package mxnet: All model parameters were initialized with a uniform distribution $U(-0.03, 0.03)$, then we applied a **stochastic gradient descent algorithm with a momentum** of 0.9, a batch-size of 50 (batch samples are randomly chosen at each iteration), and an initial learning rate of $10^{-8}$. The choice of initial learning rate was critical for a good optimization behavior. A too big learning rate can lead to training loss divergence, whereas when it is too small, learning can be very slow. We stopped when the average slope of the training mean loss had an absolute difference to 0 on the last 100 epochs inferior to $10^{-3}$. The learning took approximately 5 min for SNN, 10 min for DNN, and 5 h for CNN (independently of the version).

## 10.3.5 Evaluation Metrics

Predictions are made for every species of $E_{50}$ and several model performance metrics are calculated for each species and for two disjoints and randomly sampled subsets of sites: A train set (4781 sites) which is used for fitting all models and a test set (400 sites) which aims at testing models generalization capacities. Then, train and test metrics are averaged over the 50 species. The performance metrics are described in the following.

### 10.3.5.1 Mean Loss

Mean loss, just named loss in the following, is an important metric to consider because it is relevant regarding our ecological model and it is the objective function that is minimized during model training. The Mean loss of model $m$ on species $i$ and on sites $1, \ldots, K$ is:

$$\text{Loss}(m, i, \{1, \ldots, K\}) = \frac{1}{K} \sum_{k=1}^{K} \lambda_{m,\theta_i}(x_k) - y_k^i \log(\lambda_{m,\theta_i}(x_k))$$

In Table 10.3, the loss is averaged over species of $E_{50}$. Thus, in the case of a mono-response model, we averaged the metric over the 50 independently learned models. In the multi-response case, we averaged the metric over each species response of the same model.

### 10.3.5.2 Root Mean Square Error (Rmse)

The root mean square error is a general error measure, which, in contrary to the previous one, is independent of the statistical model:

$$\text{Rmse}(m, i, \{1, \ldots, L\}) = \sqrt{\frac{1}{K} \sum_{k=1}^{K} \left(y_k^i - \lambda_{m,\theta_i}(x_k)\right)^2}$$

In Table 10.3, the average of the **Rmse** is computed over species of $E_{50}$. Mono-response models are treated as explained previously.

### 10.3.5.3   Accuracy on 10% Densest Quadrats (A10%DQ)

It represents the proportion of sites which are in the top 10% of all sites in term of both real count and model prediction. This is a meaningful metric for many concrete scenarios where the regions of a territory have to be prioritized in terms of decision or actions related to the ecology of species. However, we have to define the last site ranked in the top 10% for real counts, which is problematic for some species, because of ex-aequo sites. That is why we defined the following procedure which adjust for each species the percentage of top cells, such that the metrics can be calculated and the percentage is the closest to 10%. Denoting $y$ the vector of real counts over sites and $\hat{y}$ the model prediction:

$$A10\%DQ(\hat{y}, y) := \frac{N_{p\&c}(\hat{y}, y)}{N_c(y)} \qquad (10.7)$$

Where $N_{p\&c}(\hat{y}, y)$ is the number of sites that are contained in the $N_c(y)$ highest values of both $y$ and $\hat{y}$.

Calculation of $N_c(y)$ : We order the sites by decreasing values of $y$ and note $C_k$ the value of the $k^{th}$ site in this order. Noting $d := \text{round}(\dim(y)/10) = \text{round}(\dim(\hat{y})/10)$, as we are interested in the sites ranked in the 10% highest, if $C_d > C_{d+1}$ we simply set $N_c(y) = d$. Otherwise, if $C_d = C_{d+1}$ (ex-aequo exist for $d^{th}$ position), we note **Sup** the position of the last site with value $C_{d+1}$ and **Inf** the position of the first site with count $C_d$. The chosen rule is to take $N_c(y)$ such that $N_c(y) = \min(|\textbf{Sup} - d|, |\textbf{Inf} - d|)$.

## 10.4   Results

In the first part we describe and comment the main results obtained from performance metrics. Then, we illustrate and discuss qualitatively the behavior of models from the comparison of their predictions maps to real counts on some species.

### 10.4.1   Quantitative Results Analysis

Table 10.3 provides the results obtained for all the evaluated models according to the three evaluation metrics. The four main conclusions that we can derive from that results are that (1) performances of LGL and mono-response DNN are lower than the one of MAXENT for all metrics, (2) multi-response DNN outperforms SNN in every version and for all metrics, (3) multi-response DNN outperforms MAXENT in test Rmse in every version, (4) CNN outperforms all the other models, in every versions (CNN50, 200, 1000), and for all metrics.

According to these results, MAXENT shows the best performance amongst mono-response models. The low performance of the baseline LGL model is mostly due to underfitting. Actually, the evaluation metrics are not better on the training set than the test set. Its simple linear architecture is not able to exploit the complex relationships between environmental variables and observed abundance. DNN shows poor results as well in the mono-response version, but for another reason. We can see that its average training loss is very close to the minimum, which shows that the model is overfitting, i.e. it adjusts too much its parameters to predict exactly the training data, loosing its generalization capacity on test data.

However, for multi-responses versions, DNN performance increases importantly. DNN50 shows better results than MAXENT for the test Loss and test Rmse, while DNN200 and DNN1000 only show better Rmse. To go deeper, we notice that average and standard deviation of test rmse across $E_{50}$ species goes down from DNN1 to DNN1000, showing that model becomes less sensitive to species data. Still, test loss and A10%DQ decrease, so there seems to be a performance trade-off between the different metrics as a side effect of the number of responses.

Whatever is the number of responses for SNN, the model is under-fitting and its performance are stable, without any big change between SNN50, 200, and 1 K. This model doesn't get improvement from the use of training data on a larger number of species. Furthermore, its performance is always lower than DNN's, which shows that stacking hidden layers improves the model capacity to extract relevant features from the environmental data, keeping all others factors constant.

The superiority of the CNN whatever the metric is a new and important result for species distribution modeling community. Something also important to notice, as for DNN, is the improvement of its performance for te.Loss and te.Rmse when the number of species in output increases. Those results suggest that the multi-response regularization is efficient when the model is complex (DNN) or the input dimensionality is important (CNN) but has no interest for simple models and small dimension input (SNN). There should be an optimal compromise to find between model complexity, in term of number of hidden layers and neurons, and the number of species set as responses.

For the best model CNN1000, it is interesting to see if the performance obtained on $E_{50}$ could be generalized at a larger taxonomic scale. Therefore, we computed the results of the CNN1000 on the 1000 plant species used in output. Metrics values are :

- Test Loss = $-1.275463$ (minimum = $-1.95$)
- Test Rmse = 2.579596
- Test A10%DQ = 0.58

These additional results show that the average performance of CNN1000 on $E_{1000}$ remains close from the one on $E_{50}$. Furthermore, one can notice the stability of performance across species. Actually, the test Rmse is lower than 3 for 710 of the 1000 species. That means that the learned environmental features are able to explain the distribution of a wide variety of species. According to the fact that French flora is compound of more than 6000 plant species, the potential of improvement of CNN

**Table 10.3** Train and test performance metrics averaged over all species of $E_{50}$ for all tested models

| # species in output | Archi. | Loss on $E_{50}$ | | Rmse on $E_{50}$ | | A10%DQ on $E_{50}$ | |
|---|---|---|---|---|---|---|---|
| | | tr.(min:-1.90) | te.(min:-1.56) | $tr.$ | te. | $tr.$ | te. |
| 1 | MAX | −1.43 | −0.862 | 2.24 | 3.18 | 0.641 | 0.548 |
| | LGL | −1.11 | −0.737 | 3.28 | 3.98 | 0.498 | 0.473 |
| | DNN | −1.62 | −0.677 | 3.00 | 3.52 | 0.741 | 0.504 |
| 50 | SNN | −1.14 | −0.710 | 3.14 | 3.05 | 0.494 | 0.460 |
| | DNN | −1.45 | −0.927 | 2.94 | 2.61 | 0.576 | 0.519 |
| | CNN | −1.82 | −0.991 | 1.18 | 2.38 | 0.846 | 0.607 |
| 200 | SNN | −1.09 | −0.690 | 3.25 | 3.03 | 0.479 | 0.447 |
| | DNN | −1.32 | −0.790 | 5.16 | 2.51 | 0.558 | 0.448 |
| | CNN | −1.59 | −1.070 | 2.04 | 2.34 | 0.650 | 0.594 |
| 1K | SNN | −1.13 | −0.724 | 3.27 | 3.03 | 0.480 | 0.455 |
| | DNN | −1.38 | −0.804 | 3.86 | 2.50 | 0.534 | 0.467 |
| | CNN | −1.70 | −1.09 | 1.51 | 2.20 | 0.736 | 0.604 |

For the single response class, the metric is averaged over the models learnt on each species

predictions based on the use of this volume of species could be really important and one of the first at the country level (which is costly in terms of time with classical approaches).

We can go a bit deeper in the understanding of model performances in terms of species types. Figure 10.3 provides for CNN1000 and MAXENT the test Rmse as a function of the species percentage of presence sites. It first illustrates the fact that all SDMs are negatively affected by an higher percentage of presence sites, even the best, which is a known issue amongst species distribution modelers. Actually, the two models have quite similar results for species with high percentage of presence sites. Moreover, CNN1000 is better for most species compared to Maxent, and especially for species with low percentage of presence sites. For those species, we also notice that CNN's variance of Rmse is much smaller than MAXENT: there is no hard failing for CNN.

## 10.4.2    Qualitative Results Analysis

As metrics are only summaries, visualization of predictions on maps can be useful to make a clearer idea of the magnitude and nature of models errors. We took a particular species with a spatially restricted distribution in France, *Festuca cinerea*, in order to illustrate some models behavior that we have found to be consistent across this kind of species in $E_{50}$. The maps of real counts and several models predictions for this species are shown on Fig. 10.4. As we can note on map A of, *Festuca cinerea* was only observed in the south east part of the French territory. When we compare the different models prediction, CNN1000 (B) is the closest to

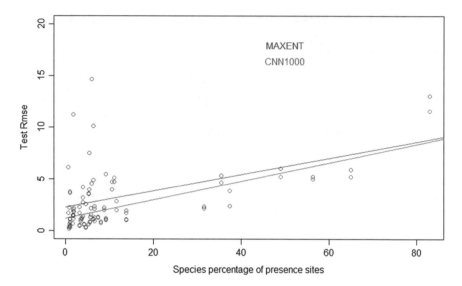

**Fig. 10.3** Test Rmse plotted versus percentage of presence sites for every species of $E_{50}$, with linear regression curve, in blue with Maxent model, in red with CNN1000

real counts though DNN50 (C) and MAXENT (E) are not far. Clearly, DNN1000 (E) and LGL (F) are the models that over estimate the most the species presence over the territory. Another thing relative to DNN behavior can be noticed regarding Fig. 10.4. DNN1000 has less peaky punctual predictions than DNN50, it looks weathered. This behavior is consistent across species and could explain that the A10%DQ metric is weak for DNN1000 (and DNN200) compared to DNN50: A contraction of predicted abundance values toward the mean will imply less risk on prediction errors but predictions on high abundance sites will be less distinguished from others.

Good results provided in Table 10.3 can hide bad behavior of the models for certain species. Indeed, when we analyze, on Fig. 10.5, the distribution predicted by Maxent and CNN1000 for widespread species, such as *Anthriscus sylvestris* (L.) and *Ranunculus repens* L., we can notice a strong divergence with the INPN data. These two species, with the most important number of observation and percentage of presence sites in our experiment (see Table 10.1), are also the less well predicted by all models. For both species, MAXENT shows very smooth variations of predictions in space, which is sharply different from their real distribution. If CNN1000 seems to better fit to the presence area, it has still a lot of errors.

As last interesting remark, we note that a global maps analysis, on more species than the ones illustrated here, shows a consistent stronger false positive ratio for models under-fitting the data or with too much regularization (high number of responses in output).

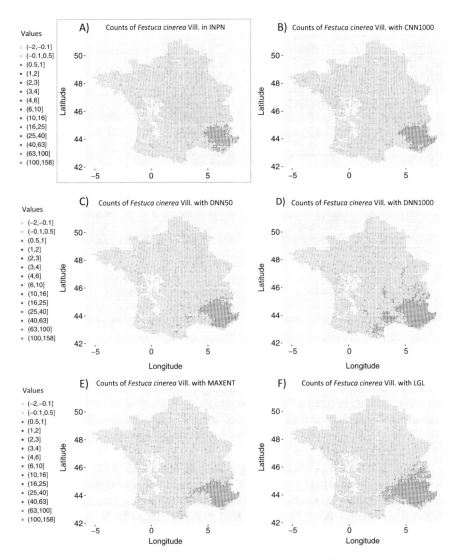

**Fig. 10.4** Real count of *Festuca cinerea* Vill. and prediction for five different models. Test sites are framed into green squares. (**a**) Number of observations in INPN dataset, and geographic distribution predicted with (**b**) CNN1000, (**c**) DNN50, (**d**) DNN1000, (**e**) Maxent, (**f**) LGL

## 10.5   Discussion

The performance increase with multi-responses models shows that multi-responses architecture are an efficient regularization scheme for NNs in SDM. It could be interesting to evaluate the performance impact of going multi-response on rare species where data rare limited. We have systematically noticed false predicted

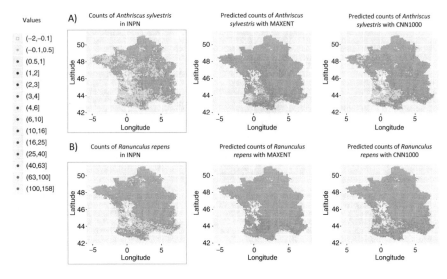

**Fig. 10.5** (**a**) Species occurrences in INPN dataset, and geographic distribution predicted with Maxent and CNN1000 for *Anthriscus sylvestris* (L.) Hoffm., (**b**) Species occurrences in INPN dataset, and geographic distribution predicted with Maxent and CNN1000 for *Ranunculus repens* L

presence for species that are not in the Mediterranean region. It could be due to a high representativity of species from this region in France. In the multi-response modeling, the Mediterranean species could favor prediction in this area through neurons activations rather than other areas where few species are present, inducing bias. Thus, the distributions complementarity between selected species could be an interesting subject for further research.

Even if our study presents promising results, there are still some open problems. A first one is related to the bias in the sampling process that is not taken into account in the model. Indeed, even if the estimation of bias in the learning process is difficult, this could strongly improve our results. Bias can be related to the facts that (1) some regions and difficult environments are clearly less inventoried than others (this can be seen with "empty region" in South western part of the country in Figs. 10.4 and 10.5); (2) some regions are much more inventoried than others, according to the human capacities of the National botanical conservatories, which have very different sizes ; (3) some common and less attractive species for naturalists are not recorded, even if they are present in prospected areas, which is a bias due to the use of opportunistic observations rather than exhaustive count data.

In the NN models learning, there is still work to be done on quick automated procedure for tuning optimization hyper-parameters, especially the initial learning rate, and we are looking for a more suited stopping rule. On the other hand, in the case of models of species distributions, we can imagine to minimize the number of not null connections in the network, to make it more interpretable, and introduce an L1-type penalty on the network parameters. This is a potential important perspective of future works.

One imperfection in our modeling approach that induces biased distribution estimate is that the representation (vector or array of environmental variables) of a site is extracted from its geographic center. MAXENT, SNN and DNN models typically only integrate the central value of the environmental variables on each site, omitting the variability within the site. Instead of that, an unbiased data generation would sample for each site many representations uniformly in its spatial domain and in number proportional to its area. This way, it would provide richer information about sites and at the same time prevent NN model over-fitting by producing more data samples.

A deeper analysis of the behavior of the models according to the ecological preferences of the species could be of a strong interest for the ecological community. This study could allow to see dependences of the models to particular spatial patterns and/or environmental variables. Plus, it would be interesting to check if NN perform better when the species environmental niche is in the intersection of variables values that are far from their typical ranges into the study domain, which is something that MAXENT cannot fit.

Another interesting perspective for this work is the fact that, new detailed fine-scale environmental data become freely available with the development of the open data movement, in particular thanks to advances in remote sensing methods. Nevertheless, as long as we only have access to spatially degraded observations data at kilometer scales like here, it is difficult to consistently estimate the effect of variables that vary at high frequency in space. For example, the informative link between species abundance and land cover, proximity to fresh water or proximity to roads, is very blurred and almost lost. To overcome this difficulty, there is much hope in the high flow of finely geolocated species observations produced by citizen sciences programs for plant biodiversity monitoring like **Tela Botanica**,[5] **iNaturalist**,[6] **Naturgucker**[7] or **Pl@ntNet**.[8] From what we can see on the **GBIF**,[9] the first three already have high resolution and large cover observation capacity: they have accumulated around three hundred thousand finely geolocated plant species observations just in France during last decade. Citizen programs in biodiversity sciences are currently developing worldwide. We expect them to reach similar volumes of observations to the sum of national museums, herbaria and conservatories in the next few years, while still maintaining a large flow of observations for the future. With good methods for dealing with sampling bias, those fine precision and large spatial scale data will make a perfect context for reaching the full potential of deep learning SDM methods. Thus, NN methods could be a significant tool to explore biodiversity data and extract new ecological knowledge in the future.

---

[5]http://www.tela-botanica.org/site:accueil.

[6]https://www.inaturalist.org/.

[7]http://naturgucker.de/enjoynature.net.

[8]https://plantnet.org/en/.

[9]https://www.gbif.org/.

## 10.6   Conclusion

This study is the first one evaluating the potential of the deep learning approach for species distributions modeling. It shows that DNN and CNN models trained on 50 plant species of French flora clearly overcomes classical approaches, such as Maxent and LGL, used in ecological studies. This result is promising for future ecological studies developed in collaboration with naturalists expert. Actually, many ecological studies are based on models that do not take into account spatial patterns in environmental variables. In this paper, we show for a random set of 50 plant species of the French flora, that CNN and DNN, when learned as multi-species output models, are able to automatically learn non-linear transformations of input environmental features that are very relevant for every species without having to think a priori about variables correlation or selection. Plus, CNN can capture extra information contained in spatial patterns of environmental variables in order to surpass other classical approaches and even DNN. We also did show that the models trained on higher number of species in output (from 50 to 1000) stabilize predictions across species or even improve them globally, according to the results that we got for several metrics used to evaluate them. This is probably one the most important outcome of our study. It opens new opportunities for the development of ecological studies based on the use of CNN and DNN (e.g. the study of communities). However, deeper investigations regarding specific conditions for models efficiency, or the limits of interpretability NN predictions should be conducted to build richer ecological models.

## References

1. Hutchinson, G. (1957). Concluding remarks. Cold spring harbor symposium on quantitative biology. 22, 415–427.
2. Hastie, T. & Tibshirani, R. (1986). Generalized Additive Models. Statistical Science, 1(3), 297–318.
3. Friedman, J. (1991). Multivariate adaptive regression splines. The annals of statistics, 1–67.
4. Phillips, S., Dudik, M., Schapire, R. (2004). A maximum entropy approach to species distribution modeling. Proceedings of the twenty-first international conference on Machine learning, 83.
5. Phillips, S., Anderson, R. & Schapire, R. (2006). Maximum entropy modeling of species geographic distributions. Ecological modelling, 190(3), 231–259.
6. Goodfellow, I., Bengio, Y. & Courville, A. (2016). Deep Learning. MIT Press.
7. Krizhevsky, A., Sutskever, I. & Hinton, G. (2012). Imagenet classification with deep convolutional neural networks. Advances in neural information processing systems, 1097–1105.
8. Lek, S., Delacoste, M., Baran, P., Dimopoulos, I., Lauga, J. & Aulagnier, S. (1996). Application of neural networks to modelling nonlinear relationships in ecology. Ecological modelling, 90(1), 39–52.
9. Thuiller, W. (2003). BIOMOD–optimizing predictions of species distributions and projecting potential future shifts under global change. Global change biology, 9(10), 1353–1362.
10. Leathwick, J.R. Elith, J. & Hastie, T. (2006). Comparative performance of generalized additive models and multivariate adaptive regression splines for statistical modelling of species distributions. Ecological modelling, 199(2), 188–196.

11. LeCun, Y. & others. (1989). Generalization and network design strategies. Connectionism in perspective, 143–155.
12. Ward, G., Hastie, T., Barry, S., Elith, J. & Leathwick, J. (2009). Presence-only data and the EM algorithm. Biometrics, 65(2), 554–563.
13. Berman, M., & Turner, T. R. (1992). Approximating point process likelihoods with GLIM. Applied Statistics, 31–38.
14. P Anderson, R., Dudk, M., Ferrier, S., Guisan, A., J Hijmans, R., Huettmann, F., ...& A Loiselle, B. (2006). Novel methods improve prediction of species' distributions from occurrence data. Ecography, 29(2), 129–151.
15. Phillips, S. & Dudik, M. (2008). Modeling of species distributions with Maxent: new extensions and a comprehensive evaluation. Ecography, 31(2), 161–175.
16. Fithian, W., & Hastie, T. (2013). Finite-sample equivalence in statistical models for presence-only data. The annals of applied statistics.7,4,1917.
17. Phillips, S. Anderson, R., Dudik, M. Schapire, R. & Blair, M. (2017). Opening the black box: an open-source release of Maxent. Ecography.
18. Rumelhart, D., Hinton, G. & Williams, R. and others (1988). Learning representations by back-propagating errors. Cognitive modeling, 5(3).
19. Nair, V. & Hinton, G. (2010). Rectified linear units improve restricted boltzmann machines. Proceedings of the 27th international conference on machine learning (ICML-10), 807–814.
20. Ioffe, S. & Szegedy, C. (2015). Batch normalization: Accelerating deep network training by reducing internal covariate shift. International Conference on Machine Learning. 448–456.
21. Dutrève, B. & Robert, S. (2016). INPN - Données flore des CBN agrégées par la FCBN. Version 1.1. SPN - Service du Patrimoine naturel, Muséum national d'Histoire naturelle, Paris. Occurrence Dataset https://doi.org/10.15468/omae84 accessed via GBIF.org on 2017-08-30.
22. Karger, D. N., Conrad, O., Bohner, J., Kawohl, T., Kreft, H., Soria-Auza, R.W. & Kessler, M. (2016). Climatologies at high resolution for the earth's land surface areas. arXiv preprint arXiv:1607.00217.
23. Karger, D. N., Conrad, O., Bohner, J., Kawohl, T., Kreft, H., Soria-Auza, R.W. & Kessler, M. (2016). CHELSEA climatologies at high resolution for the earth's land surface areas (Version 1.1).
24. Zomer, R., Bossio, D., Trabucco, A., Yuanjie, L., Gupta, D. & Singh, V. (2007). Trees and water: smallholder agroforestry on irrigated lands in Northern India.
25. Zomer, R., Trabucco, A., Bossio, D. & Verchot, L. (2008). Climate change mitigation: A spatial analysis of global land suitability for clean development mechanism afforestation and reforestation. Agriculture, ecosystems & environment, 126(1), 67–80.
26. Panagos, P. (2006). The European soil database. GEO: connexion, 5(7), 32–33.
27. Panagos, P., Van Liedekerke, M., Jones, A., Montanarella, L. (2012). European Soil Data Centre: Response to European policy support and public data requirements. Land Use Policy, 29(2),329–338.
28. Van Liedekerke, M. Jones, A. & Panagos, P. (2006). ESDBv2 Raster Library-a set of rasters derived from the European Soil Database distribution v2. 0. European Commission and the European Soil Bureau Network, CDROM, EUR, 19945.
29. Chen, T., Li, M., Li, Y., Lin, M., Wang, N., Wang, M., ...& Zhang, Z. (2015). Mxnet: A flexible and efficient machine learning library for heterogeneous distributed systems. arXiv preprint arXiv:1512.01274.

# Appendix A
# Existing Data and Metadata Standards and Schemas Related to Citizen Science

| Name | Host | Description |
|---|---|---|
| ADIwg Project Metadata Standard | The Alaska Data Integration Working Group (ADIwg) | A set of common fields for use in exchanging discovery level information about the who, what, when and where of projects in Alaska, that has been mapped to ISO 19115/19110. URL: adiwg.org |
| Biocollect | Atlas of Living Australia | Form-based structured data collection for: (1) ad-hoc survey-based records; (2) method-based systematic structured surveys; and (3) activity-based projects such as natural resource management intervention projects. It also supports upload of unstructured data in the form of data files, grey literature, images, sound bytes, videos, etc. URL: ala.org.au/biocollect |
| Darwin Core | Biodiversity Information Standards (TDWG) | A metadata specification for information about the geographic occurrence of species and the existence of specimens in collections.URL: rs.tdwg.org/dwc/index.htm |
| Data Catalog Vocabulary (DCAT) | W3C | A vocabulary that is designed to achieve interoperability between data catalogues on the web. URL: w3.org/TR/vocab-dcat |

(continued)

© Springer International Publishing AG, part of Springer Nature 2018
A. Joly et al. (eds.), *Multimedia Tools and Applications for Environmental & Biodiversity Informatics*, Multimedia Systems and Applications, https://doi.org/10.1007/978-3-319-76445-0

| Name | Host | Description |
|------|------|-------------|
| DOI | – | Digital Object Identifier: provides a system for the identification and hence management of information ("content") on digital networks, providing persistence and semantic interoperability. URL: doi.org |
| Dublin Core | The Dublin Core Metadata Initiative (DCMI) | An interoperable online metadata standard focused on networked resources. URL: dublincore.org |
| EML | – | Ecological Metadata Language is a specification developed for the ecology discipline. URL: knb.ecoinformatics.org |
| INSPIRE | EU | The EU INSPIRE Directive aims to create a Europe-wide infrastructure for public sector spatial information. By making spatial data more interoperable, it facilitates unified policies between regions, for example on the environment. To this end it specifies formats and discovery services that public authorities must use for publishing spatial data. URL: inspire.ec.europa.eu |
| ISO 19115-1:2014—Geographic information—Metadata | ISO | This metadata standard defines how to describe geographical information and associated services, including contents, spatial-temporal purchases, data quality, access and rights to use. It is maintained by the ISO/TC 211 committee. URL: iso.org/standard/53798.html and iso.org/iso/en/ CatalogueDetailPage.CatalogueDetail?CSNUMBER=26020 |
| ISO/IEC 11179 | | Describes the metadata and activities needed to manage data elements in a registry to create a common understanding of data across organizational elements and between organizations. URL: en.wikipedia.org/wiki/ISO/IEC_11179 |
| MIxS | Genomic Standards Consortium | The GSC family of minimum information standards (checklists)—Minimum Information about any (x) Sequence (MIxS) MIxS currently consists of three separate checklists; MIGS for genomes,[a] MIMS for metagenomes,[b] and MIMARKS[c] for marker genes. We created an overarching framework, the MIxS standard.[d] MIxS includes the technology-specific checklists from the previous MIGS and MIMS standards, provides a way of introducing additional checklists such as MIMARKS, and also allows annotation of sample data using environmental packages. The three checklists that are currently under MIxS share the same central set of core descriptors, but have checklist specific descriptors as well. Additionally, they enable a detailed description of environment through the use of optional environmental packages. URL: gensc.org/mixs |

(continued)

| Name | Host | Description |
|---|---|---|
| OGC SWE4CS | Open Geospatial Consortium (OGC) | Sensor Web Enablement for Citizen Science (SWE4CS) is a new standard being proposed by the Citizen Science Working Group for observations, measurements and sensing procedures as part of its standard suite to support sensor networks. URL: portal.opengeospatial.org/files/?`artifact_id`=70328 |
| Project Open Data Metadata Schema (POD) v1.1 | U.S. Government | A DCAT based vocabulary for metadata about data and APIs, as defined for federal agencies in the US. URL: project-open-data.cio.gov/v1.1/schema |
| PPSR _CORE | CitSci.org | Public Participation in Scientific Research_Core is a standard to share basic information across databases that catalog citizen science projects. It has been developed in 2013 by DataONE. URL: citsci.org/cwis438/websites/citsci/PPSR_Core_Documentation.php |

[a] wiki.gensc.org/index.php?title=MIGS/MIMS

[b] wiki.gensc.org/index.php?title=MIGS/MIMS

[c] wiki.gensc.org/index.php?title=MIMARKS

[d] Publication in Nature Biotechnology—http://www.nature.com/nbt/journal/v29/n5/full/nbt.1823.html

# Appendix B
# Creative Commons (CC) and Open Data Commons (ODC) Licenses

See Table B.1.

**Table B.1** The abbreviations in the table mean *BY* Attribution; *SA* Share-Alike; *NC* Non-Commercial; *ND* No Derivatives, ODC-PDDL: Open Data Commons Public Domain Dedication and Licence, ODC-By: Open Data Commons Attribution Licence

| License type | Abbreviation | Description |
|---|---|---|
| Attribution | CC BY and ODC-By | This license lets others distribute, remix, tweak, and build upon your work, even commercially, as long as they credit you for the original creation |
| Attribution Share Alike | CC BY-SA | This license lets others remix, tweak, and build upon your work even for commercial purposes, as long as they credit you and license their new creations under the identical terms |
| Attribution-NonCommercial | CC BY-NC | This license lets others remix, tweak, and build upon your work non-commercially, and although their new works must also acknowledge you and be non-commercial, they don't have to license their derivative works on the same terms |
| Attribution-NoDerivs | CC BY-ND | This license allows for redistribution, commercial and non-commercial, as long as it is passed along unchanged and in whole, with credit to you |
| Attribution-NonCommercial-ShareAlike | CC BY-NC-SA | This license lets others remix, tweak, and build upon your work non-commercially, as long as they credit you and license their new creations under the identical terms |
| Attribution-NonCommercial-NoDerivs | CC BY-NC-ND | This license is the most restrictive of our six main licenses, only allowing others to download your works and share them with others as long as they credit you, but they can't change them in any way or use them commercially |
| No Rights Reserved | CC0 and ODC-PDDL | Enables scientists, educators, artists and other creators and owners of copyright- or database-protected content to waive those interests in their works and thereby place them as completely as possible in the public domain, so that others may freely build upon, enhance and reuse the works for any purposes without restriction under copyright or database law |

Source: https://creativecommons.org and Groom et al. [16]

© Springer International Publishing AG, part of Springer Nature 2018
A. Joly et al. (eds.), *Multimedia Tools and Applications for Environmental & Biodiversity Informatics*, Multimedia Systems and Applications,
https://doi.org/10.1007/978-3-319-76445-0

# Appendix C
# List of Apps, Platforms and Their Functionalities in Citizen Science Projects

See Table C.1.

© Springer International Publishing AG, part of Springer Nature 2018     207
A. Joly et al. (eds.), *Multimedia Tools and Applications for Environmental & Biodiversity Informatics*, Multimedia Systems and Applications,
https://doi.org/10.1007/978-3-319-76445-0

**Table C.1** This list was put together for this chapter

| Name and date of release (tentative) | Target | Aim | (Main) functions | Scale | Holder | Link App Store and/or Play Store | Link more information |
|---|---|---|---|---|---|---|---|
| iNaturalist.org 2008 | All ages | Biodiversity monitoring | Record location and option to obscure it from the public, record photo or sound, community interaction | Worldwide | California Academy of Sciences | play.google.com/store/apps /details?id=org. inaturalist.android and itunes. apple.com/us/app/inaturalist /id421397028?mt=8 | inaturalist.org |
| Natusfera | All ages | Biodiversity monitoring | Record location, photo and sound, community interaction | Worldwide | CREA, GBIF Spain and CSIC | play.google. com/store/apps/ details?id=org. gbif.inaturalist.android | natusfera.gbif.es |
| iSpot 2008 | All ages | Biodiversity monitoring | Record location and photo, community interaction, reputation system to motivate and reward participants | Worldwide | The Open University | – | ispotnature.org |
| eBird 2002 | All ages | Bird biodiversity and abundance monitoring | Record location, date and how data were gathered including point counts, transects, and area searches. It has automated data quality filters and community interaction | Worldwide | Audubon and Cornell Lab of Ornithology | play.google.com/store/ apps/details?id=edu.cornell. birds.ebird and itunes.apple.com/us/app/ ebird-by-cornell-lab-ornithology /id988799279?mt=8 | ebird.org |
| Naturblick 2016 | All ages | Biodiversity monitoring | Record location, date, images and sounds for species identification | Local: city of Berlin | Natural History Museum Berlin | play.google.com/store/apps/ details?id=com.-mfn_berlin_stadtnatur_ entdecken.naturblick &hl=de and itunes.apple.com/de/app/ naturblick/ id1206911194?mt=8&hl=de | naturblick. naturkunde-museum. berlin/ |

| | | | | | | | |
|---|---|---|---|---|---|---|---|
| MedMIS 2013 | All ages | Marine invasive alien species (IAS) monitoring | collects location, picture and reports are displayed in a map | Regional: Mediterranean Sea | IUCN | itunes.apple. com/en/app/iucn-medmis/ id740440970?l=es& ls=1&mt=8 and play.google. com/store/apps/details? id=com.geographica.iucn _reporting&hl=en | iucn-medmis.org |
| Korina 2010 | All ages | Invasive alien plant species monitoring | collects location, picture and reports are displayed in a map | Local: Sachsony Anhalt | UfU eV. | play.google.com/store/ apps/details?id=de. korina&hl=de itunes.apple.com/de/ app/korina/ id868783957?mt=8 | korina.info |
| Invasive Alien Species in Europe, 30 November 2017 (App Store) Invasive Alien Species Europe, 20 November 2017 (Play Store) | general public (amateurs and professionals) | Informing about invasive alien species, and helping the early detection and monitoring thereof. | Receive and share information about Invasive Alien Species (IAS) in Europe.[a] The app provides details about 37 difference IAS that are considered to be of interest to the complete European Union. Users can record pictures of possible Invasive Alien Species together with complementary information about their observation. | European Union | European Commission | itunes.apple.com/it/ app/invasive-alien-species-in/ id1117811993?mt=8 and play.google. com/store/apps/ details?id=eu.europa.publications. mygeossias&hl=en | digitalearthlab. jrc.ec.europa.eu/ app/invasive-alien -species-europe |
| Loss of the Night app 2013 | 12+ (requires near and far vision) | Measure how bright the sky is by seeing how many stars are visible | Star visibility meter (via human eye observations) | Worldwide | Christopher Kyba, GFZ German Research Centre for Geosciences | play.google. com/store/apps/ details?id=com. cosalux.welovestars and https:// itunes.apple.com/en/app/loss-of-the-night/id928440562 | lossof thenight.blog spot.de/2015/01/ brief-introduction-to -loss-of-night- app.html |

(continued)

**Table C.1** (continued)

| Name and date of release (tentative) | Target | Aim | (Main) functions | Scale | Holder | Link App Store and/or Play Store | Link more information |
|---|---|---|---|---|---|---|---|
| My Sky at Night 2015 | 12+ | Improve access to light pollution data | Visualize light pollution data, evaluate accuracy of Loss of the Night app observations, plot trends in sky brightness, download data as csv files | Worl-dwide | Christo-pher Kyba, GFZ German Research Centre for Geosciences | – | myskyatnight.com |
| My Simulated Sky at Night (expected 2017) | 15+ | Improve understanding of how street light technology impacts skyglow (light pollution) | Sliders to adjust type of lighting installed, generates maps of sky brightness based on the setting | Worl-dwide | Christ-opher Kyba, GFZ German Research Centre for Geosciences | – | Not yet online |
| Fotoquest Go | All ages | Undertake 'quests' to travel to specific locations and collect information on land cover and land use based on a simplified LUCAS protocol | Stores the land cover/land use information and geotagged photographs in the Geo-Wiki database, provides gamification functions such as a leaderboard, awarding higher points for quests taken the first time, etc. | Initially applied to Austria and Europe (from Sep 2017) but could be applied Worl-dwide | Inter-national Institute for Applied Systems Analysis (IIASA) | play.google. com/store/apps/ details?id=com. IIASA.FotoQuestGo&hl=en | fotoquest-go.org |

| | | | | | | | |
|---|---|---|---|---|---|---|---|
| Picture Pile | All ages | Rapid classification of satellite images (deforestation, post-disaster damage assessment, cropland) | Simple interface for answering yes/no/maybe to one classification question by swiping image left, right or down, stores the answers to the question in the Geo-Wiki database, provides gamification functions such as leaderboards | region and case specific (disaster response, e.g. Hurricane Matthew in Haiti) but could also be applied Worldwide | International Institute for Applied Systems Analysis (IIASA) | play.google.com/store/apps/details?id=air.PicturePile and itunes.apple.com/us/app/picture-pile/id926740054?ls=1&mt=8 | geo-wiki.org/games/picturepile |
| Geo-Wiki Pictures | All ages | Take geo-tagged photographs of the landscape and classify the land cover using built-in legends or via a user-generated legend (e.g. crop types) | Stores the land cover (or customized legend attributes) and geotagged photographs in the Geo-Wiki database, geotagged photographs can be displayed and managed via the Geo-Wiki Pictures branch (www.geo-wiki.org) | Worldwide | International Institute for Applied Systems Analysis (IIASA) | play.google.com/store/apps/details?id=GeoWikiMobile.GeoWikiMobile&hl=en and itunes.apple.com/at/app/geo-wiki-mobile/id533430760 and microsoft.com/de-de/store/p/geo-wiki-picture/9nblggh4tl83 | geo-wiki.org/branches/pictures |
| LACO-Wiki Mobile | All ages | Validate land cover and land use maps on the ground | Stores land cover/land use and any geo-tagged photographs in the LACO-Wiki land cover validation repository | Can be applied at any scale from local to Worldwide | Inter-national Institute for Applied Systems Analysis (IIASA) | play.google.com/store/apps/details?id=com.geomarvel.com.lacowiki&hl=en | laco-wiki.net |

(continued)

**Table C.1** (continued)

| Name and date of release (tentative) | Target | Aim | (Main) functions | Scale | Holder | Link App Store and/or Play Store | Link more information |
|---|---|---|---|---|---|---|---|
| Hush City 2017 | All ages | Record sounds, measure the sound exposure, provide user feedback, create a collective map of "everyday quiet areas" | Audio recorder, sound levels meter | Worldwide | Antonella Radicchi, Technical University Berlin | itunes.apple.com/us/app/hush-city/id1174145857?mt=8 and play.google. com/store/apps/details? id=com.hushcity.app | opensourcesounds-capes. org/hush-city/ |
| Noise Tube 2008 | All ages | Measure the sound exposure, tag noisy sources, create a collective map of noise pollution | Sound levels meter | Worldwide | Sony Computer Science Lab in Paris (csl.sony.fr) and Software Languages Lab at the Vrije Universiteit Brussel (soft.vub .ac.be/) | play.google.com/store/apps/details?id=net.noisetube | noisetube.net/index.html#&panel1-1 |
| WideNoise 2009 | All ages | Measure the sound exposure, tag noisy sources, create a collective map of noise pollution | Sound levels meter | Worldwide | CSP, L3S Kassel / Würzburg | itunes.apple.com/app/id657693514 and play.google. com/store/apps/details?id=eu.everyaware. widenoise.android | cs.everyaware.eu/event/ widenoise/ and cs.everyaware.eu/event/widenoise |

| | | | | | | | |
|---|---|---|---|---|---|---|---|
| Ambiciti 2016 | All ages | Measure the sound exposure, develop correlation with stress levels, compute air quality, indicate the safest route to your destination | Sound levels meter, air quality calculation software | Worldwide, Paris and San Francisco (for the Air Quality maps) | Ambiciti | play.google.com/store/apps/details?id=fr.inria.mimove.quantifiedself and itunes.apple.com/us/app/ambiciti/id1080606926?mt=8 | ambiciti.io |
| Citclops/EyeOnWater 2013 | All ages | An observatory for coast and ocean optical monitoring | Several new sensor systems, based on optical technologies, to respond to a number of scientific, technical and societal objectives, ranging from more precise monitoring of key environmental descriptors of the aquatic environment (water colour, transparency and fluorescence) to an improved management of data collected with citizen participation and engagement. | Worldwide | Free | itunes.apple.com/us/app/eyeonwater-colour/id1021542366?mt=8 and play.google.com/store/apps/details?id=nl.maris.citclops.crosswalk | citclops.eu/ and eyeon-water.org/ |
| Ocean Sampling Day Citizen App (OSD-App) 2014 | All ages | Genomic Biodiversity Observations | Record location, photo and environmental data accompanying the genomic sampling event of Ocean Sampling Day | Worldwide | Micro B3 Consortium | play.google.com/store/apps/details?id=com.iw.esa&hl=en (IOS version upon request) | mb3is.megx.net/osd-app |

Take in account that such a list can become obsolete in time, however it provides references and ideas that may ease the discovery of more updated lists or tools
a easin.jrc.ec.europa.eu

# Appendix D
# Examples of Symbolic and Non-symbolic Rewards in Citizen Science Projects

| Symbolic rewards | Non-symbolic rewards |
|---|---|
| Game badges | Promotional items |
| Community badges | Prizes |
| Score on a leaderboard | Co-authorship on a scientific paper |
| Listing of top contributors | Volunteer appreciation events |
| Personal performance ratings | Payment for services |
| Naming privileges | Covering expenses that are related to the activity |
| Certificates | Scientific instruments and supplies |
| Acknowledgement through social media channels | |

© Springer International Publishing AG, part of Springer Nature 2018
A. Joly et al. (eds.), *Multimedia Tools and Applications for Environmental & Biodiversity Informatics*, Multimedia Systems and Applications, https://doi.org/10.1007/978-3-319-76445-0

# Appendix E
# List of Apps for Sonic Environment and Noise Pollution Monitoring

See Table E.1.

A. Joly et al. (eds.), *Multimedia Tools and Applications for Environmental
& Biodiversity Informatics*, Multimedia Systems and Applications,
https://doi.org/10.1007/978-3-319-76445-0

**Table E.1** List of Apps for sonic environment and noise pollution monitoring, their functionalities and links for download and discovery

| Name and date of release | Target | Aim | (Main) functions | Scale | Holder | Link App Store and/or Play Store | Link more information |
|---|---|---|---|---|---|---|---|
| Noisemap 2012 | All ages | Measure the sound exposure, tag noisy sources, create a collective map of noise pollution | Sound levels meter | Darmstadt | Technical University of Darmstadt | play.google.com/store/apps/details?id=de.tudarmstadt.tk.noisemap | tk.informatik.tu-darmstadt.de/de/research/smart-urban-networks/noisemap/ and da-sense.de |
| I-SAY Sound Around You 2012 | All ages | Record sounds from the environment, tag soundscapes, create a collective sound map | Audio recorder | UK, Worldwide | The Audio and Acoustic Engineering Research Centre,[a] University of Salford (UK) | apple.com/itunes/download/ | soundaroundyou.com |
| Soundscape Characterization Tool 2013 | All ages | Record soundscapes, tag soundscapes | Audio recorder | Worldwide | Per Hedfors, Swedish University of Agricultural Sciences (SLU) Uppsala | itunes.apple.com/us/app/soundscape-characterization-tool/id773077497?mt=8 | – |
| Radio aporee 2013 | All ages | Record soundscapes, tag soundscapes, create a collective real time sound map | Audio recorder | Worldwide | Udo Noll | itunes.apple.com/de/app/radio-aporee/id640921893?mt=8 and play.google.com/store/apps/details?id=com.aporee.radio&hl=it | aporee.org/maps |

| | | | | | | | |
|---|---|---|---|---|---|---|---|
| stereopublic 2013 | All ages | Record soundscapes, create ambient compositions, create a collective sound art map of quietness | Audio recorder | Worldwide | Jason Sweeney | – | quiet-ecology.com/public |
| Cart_ASUR 2014 | All ages | Measure the sound exposure, tag noisy sources, create a collective map of noise pollution | Sound levels meter | Paris | MRTE team, University of Cergy Pontoise and Brus-Sense team, Vrije Universiteit Brussel | windowsphoneapks.com/APK_Cart-ASUR-NoiseTube-Mobile_Windows -Phone.html | noisetube.net/cartasur#&panel1-1 |
| Geluidenjager 2014 | All ages | Record soundscapes, tag soundscapes, create a collective sound map | Audio recorder | The Nether-lands | Neder- lands Instituut voor Beeld en Geluid | play.google.com/store/apps/details? id=net.webmapper. gvnl and itunes.apple.com/ca/app/geluidenjager /id530062802?mt=8 | geluidvannederland.nl |
| Recho 2014 | All ages | Record soundscapes and narratives, create a hidden sound map | Audio recorder | Worldwide | Recho ApS | itunes.apple.com/us/app/recho/id865541527? mt=8 | recho.org |
| Record the Earth 2014 | All ages | Record sounds of the earth, create a collective sound map | Audio recorder | Worldwide | Purdue University | play.google.com/store/apps/details?id= com.recordtheearth and itunes.apple.com/us/app/soundscape-recorder/id836741158 ?mt=8 | recordtheearth.org |

(continued)

**Table E.1** (continued)

| Name and date of release | Target | Aim | (Main) functions | Scale | Holder | Link App Store and/or Play Store | Link more information |
|---|---|---|---|---|---|---|---|
| The Noise App 2015 | All ages | Measure the sound exposure, report complaints | Sound levels meter | World-dwide | Noise Nouisance | itunes.apple.com/gb/app/the-noise-app/id926445612?mt=8 and play.google.com/store/apps/details?id=com.rhe.noiseapp &hl=en_GB | noisenuisance.org/the-app/ and thenoiseapp.com |
| Sound City 2015 | All ages | Measure the sound exposure, develop correlation with stress levels | Sound levels meter | World-dwide | Inria at Silicon valley | This app turned into Ambiciti (see Table C.1) | urbancivics.com/soundc-ity_app.html |
| Aircasting 2016 | All ages | Record, map, and share: sound levels, temperature, humidity, and fine particulate matter (PM2.5), CO, NO2 gas, heart rate, heart rate variability, R to R, breathing rate, activity level, peak acceleration, and core temperature | Sound levels meter, Arduino-powered AirBeam, Arduino-powered AirCasting Air Monitor, Zephyr BioHarness 3, Zephyr HxM | World-dwide | Habitat- map | play.google.com/store/apps/details?id =pl.llp.aircasting &hl=en | aircasting.org and habitatmap.org |

| Think About Sound 2015 | All ages | Record sounds, rate the sounds and their impact on user feelings, create a collective 3D sound map | Audio record | UK | Adam Craig, Glasgow Caledonian University | itunes.apple.com/gb/app/think-about-sound/id969517179?mt=8 and play.google.com/store/apps/details?id=com.adamcraig.thinkaboutsound | glasgow3dsoundmap.co.uk/soundmap.html |
|---|---|---|---|---|---|---|---|
| City Soundscape 2016 | All ages | Measure the sound exposure, provide user feedback | Sound levels meter | Puglia (Italy) | Alba Project s.r.l. | play.google.com/store/apps/details?id=it.albaproject.citysoundscape&hl=it | citysoundscape.it/ and fi-frontiercities.eu/#!City-Soundscape-signs-threeyear-convention-with-ASSTRA—Puglia-Transport-Association/gdbtt/5731a8ba0cf238e05b83deee |
| MoSart 2016 | All ages | Record sounds, appraise the sounds of the environment | Audio Recorder | Worldwide | Sound-Appraisal | itunes.apple.com/us/app/mosart/id1110652343 and play.google.com/store/apps/details?id=org.auditoryenvironments.mosart | soundappraisal.eu/soundapp.html |

Source: Radicchi (in press) [62]
[a] acoustics.salford.ac.uk

Printed in the United States
By Bookmasters